THEORETICAL
PHYSICS
FOR THE MASSES

Popular Physics for the General Reader

More Than Curious: A Science Memoir
by William Henry Press
ISBN: 978-981-98-0176-3
ISBN: 978-981-98-0300-2 (pbk)

No Wisdom without Folly:
The Extraordinary Life of François Englert, Nobel Laureate
by Danielle Losman
ISBN: 978-981-12-8324-6
ISBN: 978-981-12-8391-8 (pbk)

Newton . Faraday . Einstein: From Classical Physics to Modern Physics
by Tadayoshi Shioyama
ISBN: 978-981-12-3567-2
ISBN: 978-981-12-3624-2 (pbk)

Forks in the Road: A Life in Physics
by Stanley Deser
ISBN: 978-981-12-3418-7
ISBN: 978-981-12-3566-5 (pbk)

Chasing the Ghost: Nobelist Fred Reines and the Neutrino
by Leonard A Cole
ISBN: 978-981-12-3105-6
ISBN: 978-981-12-3148-3 (pbk)

The HERMES Experiment: A Personal Story
by Richard Milner and Erhard Steffens
ISBN: 978-981-12-1533-9

THEORETICAL PHYSICS
FOR THE MASSES

Brent J. Lewis

Royal Military College of Canada, Canada

W🌐 World Scientific

NEW JERSEY · LONDON · SINGAPORE · BEIJING · SHANGHAI · HONG KONG · TAIPEI · CHENNAI

Published by

World Scientific Publishing Co. Pte. Ltd.

5 Toh Tuck Link, Singapore 596224

USA office: 27 Warren Street, Suite 401-402, Hackensack, NJ 07601

UK office: 57 Shelton Street, Covent Garden, London WC2H 9HE

Library of Congress Control Number: 2024043680

British Library Cataloguing-in-Publication Data
A catalogue record for this book is available from the British Library.

THEORETICAL PHYSICS FOR THE MASSES

ISBN 978-981-98-0030-8 (hardcover)
ISBN 978-981-98-0031-5 (ebook for institutions)
ISBN 978-981-98-0032-2 (ebook for individuals)

For any available supplementary material, please visit
https://www.worldscientific.com/worldscibooks/10.1142/14028#t=suppl

Typeset by Stallion Press
Email: enquiries@stallionpress.com

Preface

This book was inspired from George Gamow's book *One, Two, Three...
Infinity, Facts and Speculations of Science*, written over some 60 years
ago. Gamow was an early advocate and pioneer of the Big Bang theory
in cosmology. His work has inspired generations of young readers in their
pursuit of more general knowledge in science. Other important books in
the foundation of theoretical physics, with a focus on the story of the
creation and expansion of the universe for the layperson, include: Steven
Weinberg's, *The First Three Minutes, A Modern View of the Origin of
the Universe*. Weinberg himself was a Nobel laureate for his authoritative
work on the unification of the weak force and electromagnetic interaction
of elementary particles. *A Brief History of Time*, along with other best-
selling books by the cosmologist Steven Hawking, are popular accounts
that capture the imagination of the general reader. Hawking contributed
to a better appreciation of the physics of black holes. Steven Gubser,
following his inspirational work to relate quantum gravity and quantum
field theories, provided a concise insight into the main ideas of string theory
as a possible "theory of everything" with his book *The Little Book of String
Theory*. John Moffat has also written a thought-provoking book on an
alternative gravitational theory entitled *Reinventing Gravity*. This book
offers challenging ideas for debate with the proposal of a modified gravity
theory to possibly eliminate the concept of dark matter that has yet to
be detected but required to explain astrophysical observations of galactic
rotations. These general accounts have helped to broaden the knowledge
and interest in science at large.

The first part of the book provides a history and collection of modern
theories of our universe. It is presented in a non-mathematical format with
no equations. Thus, the first part is completely standalone that can be easily

read separately for general interest of recent developments in theoretical physics over the past century.

However, the supplemental mathematics sections in the second part of the book provide for a more in-depth treatment with the derivation of relevant equations. It allows for the mathematically-inclined reader to better appreciate the general content covered in the first part. It provides answers to several questions such as, "how were such theories developed and what is their fundamental theoretical basis"? The second part of the book provides some supplemental mathematical foundations of theoretical physics in the same subject areas as the first part. It may be of particular interest to physicists planning on-going advanced studies. Moreover, it may attract interest from physicists, engineers and scientists in other disciplines who have always had a curiosity to better understand the important foundations of modern physics.

This treatise covers a general overview of the foundations of classical physics, special and general relativity theory, quantum mechanics and field theory including loop quantum gravity, the standard model of particle physics, string theory, cosmology theory detailing the evolution of the universe, and black hole theory. The subject matter has been selected to bring forth the most interesting and evolutionary ideas of theoretical physics. The detailed supplemental derivations afford a deeper understanding of the subject material. For instance, key equations are derived to test general relativity against classical tests as a falsifiable theory. Mathematical approximations of general relativity are shown to yield Newtonian gravitational theory as well as how gravitational waves arise from the general field equations. It describes how quantum field theory can account for particle transformation reactions. The continuous quantum field is shown to be composed of an infinite number of quantum harmonic oscillators, where particles are in fact excitations of the underlying quantum field. This work lays the groundwork for the standard model of particle physics. It provides for a classification of observed elementary particles and for the presence of three of the four fundamental forces of nature. It shows how force interactions are manifested with an exchange of virtual particles. It explains mathematically how the Higgs mechanism, through the Higgs field and associated boson, give rise to the mass of elementary particles with spontaneous symmetry breaking. Equations are further derived to explain the required number of extra dimensions as well as their compactification for a physical representation of string theory. The theoretical treatments also show how general relativity can be used to predict the cosmological equations that describe the expansion of the universe as well as black hole fundamental characteristics. Loop quantum gravity couples quantum mechanics with general relativity in an effort to resolve singularities that arise when general relativity breaks down especially for a description of the very early universe.

This book therefore provides for an interesting journey with a broad overview of theoretical physics to better understand the evolving universe that we live in.

Acknowledgements

The author would like to thank Viktor Toth for helpful comments on the book. I would also like to thank the Editor, Soh Yong Qi, and the various staff at World Scientific Publishing, for their invaluable assistance.

To my wife, Patricia and daughters, Kimberley and Jessica

Contents

Chapter 1

Introduction

1.1 History of Modern Theoretical Physics

Physical laws of nature are realized through experimental study and observation. Such laws may be captured or explained by theoretical physics with the development of mathematical models. The 20th century gave birth to modern theoretical physics. Although thermodynamics, kinetic theory, as well as conservation laws of mass, energy and momentum had already been well established, "Maxwell's equations" published in 1864 were able to detail how electric charges/currents create electric and magnetic fields. In particular, these equations captured "Gauss' law" for both electric and magnetic fields, "Faraday's law" for the induction of an electric field from a changing magnetic field, and "Ampere's law" for the relationship between electric currents and the generated magnetic field. The "Lorentz force law" described the movement of a charged particle in an electric and magnetic field. These equations together formed the foundation of classical electromagnetism, optics and electric circuits.

However, some doubts arose at the beginning of the 20th century concerning the completeness of some classical treatments to explain observed phenomena. For instance, the emission of electrons when light shines on a metal surface due to the absorption of electromagnetic radiation in a material led to an unexplained "photoelectric effect". An "ultraviolet catastrophe" was also predicted using classical physics theory for a "black body". The black body is a perfect absorber of light, which does not reflect any light. Radiation that hits it is absorbed and then re-emitted as thermal radiation (which is a function of temperature). Thus, a black body absorbs all colours of light and emits radiation. It can be represented by a cavity that contains a small hole in it where the absorptive walls of the body are painted black. Consequently, this body will absorb all incident

1

electromagnetic radiation regardless of the frequency or incidence of the radiation. However, at thermal equilibrium, the classical theory predicted that it would emit an unbounded amount of light per unit area per unit time (i.e., spectral radiance) as the wavelength of the light became shorter towards the ultraviolet end of the spectrum. This prediction was contrary to observation and required a revolutionary new theory to replace the classical one.

In 1900, Max Planck showed that electromagnetic radiation is instead emitted or absorbed in discrete packets as a "quanta" of energy. The Planck's law simply states that the energy of the radiation is proportional to its frequency. The proportionality constant is now termed the "Planck's constant". This embodied law yielded a corrected theoretical spectral distribution that solved the black body problem. It further led to the birth of "quantum physics". Albert Einstein and Satyendra Nath Bose postulated that these quanta were in fact real physical particles (now popularly called "photons"). Bose also deduced that the photons are identical in nature, giving rise to so-called "Bose–Einstein statistics", which is an important phenomenon in the description of particle physics behaviour. The existence of these quanta was further able to explain the "photoelectric effect" for which Einstein received a Nobel prize in 1921.

Important accomplishments of modern theoretical physics arose with resultant revolutionary theories of "relativity" and "quantum mechanics". The theory of the motion of bodies described by Newtonian kinematics was previously formulated with an absolute frame of reference. An absolute frame of reference is considered to be at rest with respect to the "fixed stars", which are far away and do not affect motions of nearby objects. This frame of reference thereby provides a fixed point of reference for making measurements of position and time. This concept, however, was replaced by the "special theory of relativity" that showed that an absolute frame of reference does not exist in the universe. The theory of gravity developed by Isaac Newton was succeeded with the more comprehensive "general theory of relativity" as proposed by Einstein. As previously explained, the beginnings of quantum mechanics led to a solution for the prediction of the observed spectral radiance from a black body. It also provided a better understanding of anomalies that arose in the description of specific heats of solids as well as for the internal structure of atoms and molecules. Quantum mechanics further paved the way for the development of quantum field theory in the late 1920s. Quantum field theory was able to provide a better understanding of the phenomena of superconductivity and phase transitions as well as a description of condensed matter physics. It also led to the formulation of the standard model of particle physics involving quark theory in the 1960s and 1970s. In addition, theoretical physics led to further applications of relativity to the field of cosmology for an understanding and study of the universe at large.

In the special theory of relativity of 1905, the laws of physics do not change, that is, they are "invariant" in all "inertial frames of reference". In addition, the speed of light is importantly recognized as a constant. An inertial frame of reference is one that is not undergoing any acceleration in which there is no force acting on an object that is moving at a constant velocity. The concept of length and time therefore vary relative to how an observer and object are moving with respect to one another. In relativity theory, measurements in one inertial frame can be transformed to provide measurements in another frame of reference with the use of a "Lorentz transformation". This transformation relates space and time coordinates of two systems moving at a constant velocity relative to one another. The key property of this transformation is that it preserves the space–time interval between any two events. The Lorentz transformation law is different from that of a simple "Galilean transformation", which uses an absolute frame of reference as applied in Newtonian physics. A Galilean transformation approximates a Lorentz transformation for velocities much lower than the speed of light.

An aether permeating throughout space had originally been postulated as a necessary type of medium to carry light waves. Relativity showed that there was no physical basis for electromagnetic wave propagation through a hypothetical "luminiferous aether". In fact, the Lorentz transformation used in special relativity was consistent with the null result of the Michelson-Morley experiment performed in 1887. This experiment did not reveal a different speed of light in different directions as the Earth moved through the aether wind. This result demonstrated that the speed of light was constant in any frame. Moreover, for objects moving close to the speed of light, special relativity shows that clocks associated with those objects will run more slowly, and that objects will shorten in length, compared to measurements made by an observer on Earth. This theory also proposes an equivalence between energy and mass.

In 1915, relativity was extended further to account for non-uniform acceleration. The curvature of so-called "space–time" was a mathematical concept used to replace Newton's universal law of gravity. Space–time is a single concept proposed by Hermann Minkowski in 1908 with the coupling of three space dimensions and that of time into a four-dimensional "topological space" as a means to reformulate Einstein's special theory of relativity. In general relativity theory, the presence of mass causes a curvature of space–time. Einstein demonstrated the equivalence between the inertial force of acceleration and the force of gravity. The curvature of space–time specifically restricts the path that freely-falling objects follow. In addition, general relativity predicted the bending of light around the sun and led to the prediction of black holes with the solution of his "general field equations".

Symmetry is an important concept and property that arises in theoretical physics that allows a change in perspective. Symmetries can be external (i.e., global) in nature that depend on changes of space–time as a whole. There also exist internal symmetries that do not involve any changes with respect to space–time. A famous theorem in classical physics is known as "Noether's theorem", which was proposed in the same year as general relativity. This theorem was named after its discoverer Emmy Noether, a German mathematician. It allows one to relate symmetries to conserved quantities such as charge, energy and momentum, which are basic quantities of physics. Such symmetries can result in variations of the so-called "field". The field represents some physical quantity that is assigned to every point in space–time (for example, an electric or magnetic field). The symmetries can also specifically affect a so-called "Lagrangian function" (or "Lagrangian") that is used to characterize the state of a physical system in terms of the kinetic energy of motion minus the potential energy of position. The Lagrangian formulation of a physical system provides a mathematical means to describe the dynamics of a system. A symmetry, in fact, allows the equations of motion to remain unchanged. Noether's theorem has played an important role in the development of mathematical theoretical physics.

Quantum mechanics deals with the behaviour of atoms at the subatomic scale. It solved the ultraviolet catastrophe problem for black body radiation. The basic ideas of this theory were developed by Max Planck in 1900. Quantum mechanics was also in agreement with the observed "Compton effect". This effect demonstrated that light carries momentum in which a photon can be scattered after interaction with a stationary electron in an atom. Part of the energy of the photon is then transferred to an ejected electron.

In addition, as shown by Louis de Broglie, electromagnetic light exhibits a "wave-particle duality". A full scale theory was developed in the 1920s that expressed a probabilistic nature between discrete quantum states.

This formulation was captured with important contributions by Werner Heisenberg, Max Born, Wolfgang Pauli, Paul Dirac and Erwin Schrödinger. Importantly, an "uncertainty principle", proposed by Heisenberg in 1927, indicated that there is an uncertainty in the fundamental limit for the product of the accuracy of certain physical quantities, such as momentum and position of a particle in relation to Planck's constant. The interpretation of quantum mechanics, attributed to Niels Bohr and Werner Heisenberg, became known as the "Copenhagen interpretation". This interpretation reflected on how an observer and system that is being observed are affected as well as how irreversible processes occur within a quantum mechanical system.

In quantum mechanics, the "spin" of a particle is a type of intrinsic angular momentum. The statistical properties of particle spin is such that particles with a half-integer spin (i.e., fermions) obey "Fermi–Dirac statistics", while particles with integer spin (i.e., bosons) obey "Bose–Einstein statistics". Fundamental bosons are seen to transmit forces of nature. Fermions, on the other hand, are the usual constituents of matter including electrons and nucleons (e.g., protons and neutrons). In Fermi–Dirac statistics, no two particles can occupy the same quantum state simultaneously but must occupy different energy levels. This property is known as the "Pauli exclusion principle". In comparison, in Bose–Einstein statistics, multiple particles can occupy the same quantum state simultaneously and therefore "condense" into the same lowest-energy state. This difference yields different thermodynamic properties and behaviours. For example, photons transmit the force of electromagnetism. In condensed matter physics, for a gas of bosons at very low densities that are cooled to temperatures near absolute zero, Bose–Einstein statistics are able to describe this lowest quantum state of matter.

Quantum mechanics continued to be developed, including a relativistic quantum theory by Paul Dirac in 1928 that led to the prediction for the existence of antimatter. However, infinities for energies arose with the quantization of electromagnetic theory. A technique of "renormalization" ensued, as independently developed by Julian Schwinger, Richard Feynman and Sin-Itiro Tomoonaga after the end of World War II. This approach led to a more robust theory of "quantum electrodynamics". Moreover, the concept of "quantization of fields" arose through "exchange forces" with the exchange of short-lived virtual particles as permitted within the uncertainty laws formulated for quantum mechanics. For instance, the "pion", with a mass between an electron and proton, was identified in 1947 and recognized by Hideki Yukawa as the particle that mediated the short-range strong force that kept the nucleus together.

1.1.1 *Discovery of subatomic particles*

Historically, the study of fields precluded particle physics where, at the end of the 1800's, several types of rays had been discovered. Cathode rays, which are now known as accelerated beams of electrons to energies around 10^5 eV, were first observed in 1869 by Hittorf and Plücker marking the beginning of modern day particle physics. In contrast, some 150 years later, the large electron-positron collider (LPC) at CERN, accelerates electrons to energies around 2×10^{11} eV, while the large hadron collider (LHC) accelerates protons to energies around 6×10^{12} eV.

Using cathode rays in 1895, Wilhelm Röntgen discovered X-rays. Henri Becquerel using uranium salts in 1896 created silhouetted images on covered photographic plates. Marie and Pierre Curie then discovered other radioactive elements of thorium, polonium and radium. Ernest Rutherford realized that uranium had emitted two different types of radiation called α-rays and β-rays. In 1897, J.J. Thompson measured the deflection of cathode rays by electric and magnetic fields and thereby determined the ratio of the electric charge-to-mass ratio of the electron. By 1899, he was further able to uniquely determine the electric charge using a cloud chamber. A more precise method developed by Millikan and Fletcher in 1909 using oil droplets, instead of a cloud chamber, was used to similarly balance gravitational and electric forces where the electric charge was evaluated within 1% of the modern value.

Hans Geiger and Ernest Marsden carried out an experiment by firing alpha particles at a thin foil and measuring their deflection. As an interpretation of the experimental results, Rutherford postulated in 1911 that each atom contains a heavy nucleus at the centre with a positive charge Q. With this explanation, elements were labelled by their atomic weight A and Z that was identified as the positive charge of the nucleus. Rutherford developed a scattering formula that provided an additional means to assess the charge Q of the nucleus leading to an ordering of elements. Rutherford also bombarded nitrogen by alpha particles demonstrating that a hydrogen nucleus was omitted. This particle was named the proton. Hence, by 1911 the nucleus was discovered, with the existence of the proton by 1920. In 1932, James Chadwick announced that the nucleus also contained an uncharged particle called the neutron.

In the decade between these discoveries, quantum physics was formulated with the basis of quantum field theory outlined. In this formulation, all fields associated with spin-1/2 matter particles were described by an equation developed by Dirac in 1928 that was consistent with the principles of quantum mechanics and special relativity. It was originally developed to describe the property of electrons and its solution led to the discovery of anti-matter in 1931 with the identification of positively-charged electrons called positrons. In 1932, Carl Anderson experimentally discovered the positron from cosmic ray studies with a cloud chamber.

Other particles were subsequently found from 1934 to 1937 with the examination of cosmic ray tracks. These particles were termed "muons" and "mesons". Muons and mesons are both a type of subatomic particle but belong to different categories with distinct properties. Muons are elementary particles similar to electrons but heavier. They are fundamental particles (leptons) with a single type (plus antiparticle) involved in weak interactions commonly found in cosmic ray showers. Mesons are composite particles (hadrons) made up of quark-antiquark pairs. Typical examples of mesons include pions (π^+, π^0 and π^-) and kaons (K^+, K^0 and K^-). Mesons are produced in high-energy collisions, which are unstable and decay into lighter particles with half-lives that vary from fractions of a second to relatively long-lived particles. Mesons play a crucial role in mediating the strong nuclear force between nucleons (protons and neutrons) with various types involved in particle interactions and decay processes.

In 1934, Hideki Yukawa suggested the existence of a new particle, which was later realized to bind neutrons and protons together in the nucleus as a manifestation of the strong force. By 1939, a new particle, identified as a "meson", was identified that had a mass 200 times heavier than the electron. However, experiments in 1946 showed that the interaction of the new meson was 10^{12} times weaker than that predicted by Yukawa's theory. However, in 1947, Cecil Powell discovered the "pion" by coating a glass plate with a photographic emulsion and exposing these plates at high altitudes to cosmic rays in balloons and on mountains. This particle had the properties expected by Yukawa's meson.

Later in 1947, extensive cosmic rays studies led to the identification of other particles with masses between 770 and 1600 times that of the electron. These particles were dubbed "V-particles" because of the V-shaped tracks they left behind. The V-particles were subsequently named "kaons". The next decade focussed on the classification of a plethora of particles found. The discovery of V-particles marked the beginning of a new era in particle physics. By the mid-1950's accelerator studies were competitive with cosmic ray studies where one was able to produce man-made muons, pions and kaons.

More recently, important discoveries included the top quark in 1995, the tau neutrino in 2000 and the Higgs boson in 2012.

1.1.2 *Unified field theories*

As mentioned, a unified theory for electromagnetism was developed by Maxwell in 1865 as "classical electrodynamics". In 1967, Sheldon Glashow, Abdus Salam and Steven Weinberg unified electromagnetism with the weak force as "electroweak theory". The weak force describes radioactive beta decay in which a neutron decays into a proton, electron and anti-neutrino. Particles which are carriers of the weak force as described in Section 4.1 include: W^+, W^- and Z^0. These particles were observed in accelerator experiments in 1983. As mentioned, the photon is the specific carrier of the electromagnetic force. Four forces are known to exist in nature:

(i) Gravity (first described by Newton and then by Einstein as the general theory of relativity as a curvature of dynamical space–time);
(ii) Electromagnetic force (described by Maxwell's equations);
(iii) Weak force (responsible for radioactive beta decay);
(iv) Strong force, (nowadays called the colour force, which holds together the constituents (i.e., quarks) of the neutron, proton, pions and other subnuclear particles).

A comprehensive theory of physics is expected to include the "Standard Model" of particle physics, as detailed in Chapter 3, which provides a complete classification of all elementary particles along with a unification of the electroweak and strong forces. In the 1970s, so-called "Grand Unified Theories (GUTs)" were introduced with a prediction of proton decay on average after about 10^{32} years; however, observations as early as 2009 suggested much greater proton lifetimes. As such, an ad hoc theory now unifies the electroweak and strong force but where these forces act separately in the Standard Model so that they are not truly unified. Also a more complete version of the Standard Model may include "Supersymmetry" which relates the bosons (force carriers) to the fermions (matter particles), i.e., this symmetry provides a basis for a quantum theory of gravity called "supergravity" since the general theory of relativity is a classical model. It needs to be included into the particle physics framework of the Standard Model (which is a quantum theory) to account for the physics very near the time of the Big Bang and for that of black holes.

"String Theory" has been suggested as a possible candidate for a unified theory of all forces in nature, where particles are not points but rather vibration patterns of a finite-length string with no height or width. In 1994, it was discovered that five different string theories and supergravity are approximations to a more fundamental theory, called M theory. Extra internal dimensions in this theory are not observed in the normal day-to-day physical world. Instead, these dimensions are tightly curled up where the internal space determines the value of physical constants such as the charge of the electron and the nature of interactions between elementary particles.

The Higgs boson is an elementary particle in the Standard Model of particle physics. It was predicted by Peter Higgs and several other teams in 1964 who theoretically proposed a mechanism that explained its existence. The Higgs boson was detected in experiments at the Large Hadron Collider in 2012 and confirmed on its 10th-year anniversary of discovery. Its detection supports the existence of a hypothetical Higgs field, which is the simplest of several proposed mechanisms for the spontaneous symmetry breaking of the electroweak field. This mechanism provides the means by which elementary particles in the Standard Model acquire mass. The leading explanation for this process is that there exists a field that has a non-zero strength everywhere even in otherwise empty space. Particles acquire mass with their interaction with this field. The matching particle, which is the smallest possible excitation of the Higgs field, is a crucial accomplishment for the validation of this theory.

Neutrinos are subatomic particles very similar to an electron but have no electrical charge and a very small mass (which was originally thought to be zero). These particles come in three different types (flavours), which can oscillate and switch identities from one flavour to another as demonstrated in recent experiments. Importantly, these experiments have dispelled the long held notion that neutrinos are massless — an observation which is in conflict with the Standard Model of particle physics. Neutrinos may have an important effect on the evolution of the universe and how much mass it contains. "Dark matter", which does not reflect, absorb nor emit electromagnetic radiation, is the missing matter that explains the flat rotation curves in galaxies. In physical cosmology, "dark energy" causes the accelerated expansion of the universe (i.e, a gravitationally-repulsive negative mass). The presence of dark matter affects the gravitational pull of clusters of galaxies.

MOdified Gravity (MOG) theories have been proposed to alternatively interpret the rotational dynamics of galaxies and the cluster of galaxies without the need for dark matter that has not yet been identified. These theories include, for instance: "modified Newtonian dynamics", "tensor-vector-scalar gravity" and "entropic gravity". These theories are direct competition to the well-tested theory of general relativity and, as such, have not gained wide acceptance among the scientific community.

1.2   Organization of Book

The book provides a non-mathematical description of theoretical physics in the first part for readers who wish to understand the achievements of modern theoretical physics at an overview level. It aims to provide an explanation of a number of important revolutionary theories that arose over the many years. The second supplemental part of the book provides the theoretical underpinnings with a more detailed mathematical description.

An overview of the history of theoretical physics is given in Chapter 1, beginning with the evolution of relativity and quantum mechanics. With the emergence of relativity, Newtonian classical mechanics and gravity were replaced by special relativity and the more general theory of relativity as detailed in Chapter 2. The book also shows in Chapter 3 how quantum mechanics can be used to explain behaviour at a scale smaller than the atom. This latter chapter further explains the evolution of quantum field theory as a theoretical approach to explain the interaction of subatomic particles with a combination of classical field theory, special relatively and quantum mechanics. Quantum field theory itself has formed the basis of the current Standard Model of particle physics. Loop quantum gravity theory based on a merging of quantum mechanics with general relativity can incorporate matter within the framework of quantum gravity. This latter theory competes with string theory as a way to study the quantum properties of the gravitational field and geometry of space–time. The Standard Model in Chapter 4 accounts for the behaviour of the electromagnetic, strong, and weak forces of nature through the exchange of particles and the classification of elementary particles as observed in accelerator experiments. In Chapter 5, string theory describes an approach to unify all known forces in nature. The application of these developments to the field of cosmology in Chapter 6 accounts for the nature of the expanding universe. Moreover, Chapter 7 explains how general relativity can be used to predict the behaviour of black holes, where gravity is so strong that particles or electromagnetic radiation cannot escape from this region. Theoretical physics yet to be resolved is summarized in Chapter 8.

For completeness, underlying mathematical developments of these theories are presented as a separate addendum in seven complementary appendices (i.e., Appendices A–G) for the interested reader. This addendum covers the same information as presented in a more general context in Chapters 1–7. The added supplemental material, containing key mathematical derivations and equations, is not needed to achieve an overall understanding of the subject but provides a means to better understand the basis of the theories. A glossary of terms is also given in Appendix H.

Chapter 2

Relativity

The theory of relativity considers the special theory developed in 1905. This theory, as described in Section 2.1, applies to physical phenomena without gravity. The general theory proposed in 1915, as detailed in Section 2.2, further explains the laws of gravity. Relativity helped to transform the basis of theoretical physics. It superseded Sir Isaac Newton's theory of mechanics detailing the laws of motion of macroscopic and astronomical objects as well as classical gravitational theory.

2.1 Special Relativity

The special theory was described in Einstein's 1905 paper "On the Electrodynamics of Moving Bodies". The basis of this theory considers three important postulates, which differ from the classical theory:

(i) *The principle of relativity.* The laws of physics remain the same in all inertial reference frames.

(ii) *The speed of light is constant (i.e., invariant).* All observers in inertial frames will measure the same speed of light in vacuum regardless of their relative motion or of the motion of the light source itself.

(iii) *Uniform motion is invariant.* A particle that is at rest or has a constant velocity in one inertial frame will be at rest or have constant velocity in all inertial frames.

The postulate in (ii) explained the Michelson-Morley experimental results. This experiment was an attempt to detect the presence of an "aether" that was thought to permeate throughout all space and importantly act as a carrier of light waves. The experiment was designed to detect any difference in the speed of light in the direction of movement through the aether and at right angles. However, no measured difference was found. This observation provided strong evidence against the concept of a luminiferous aether wind and gave support for the second postulate of special relativity.

The special theory of relativity importantly replaced the concept of a "Galilean transformation" as considered in classical Newtonian mechanics with a so-called "Lorentz transformation". A Galilean transformation provided a means to transform from one reference frame to another in which the two frames of reference differ only by a constant relative motion. On the other hand, the Lorentz transformation is more general in nature. It represents instead a six-parameter family of linear transformations from one coordinate frame to another that differ by a relative constant velocity. However, a Lorentz transformation is described in terms of both a physical three-dimensional space but also coupled with time through the introduction of a combined quantity known as "space–time". As such, a particular event is something that happens at a point in space and at an instant of time.

An important concept is the path that an object traces out in space–time. This path is more complicated than such concepts as an orbit or trajectory since the path also encompasses a time dimension. The path through four-dimensional space–time is known as a "world line". Of further interest in both special and general relativity is the path of a light ray. This path is called a "light cone". The light cone details how a flash of light will emanate from a single event localized at a single point in both time and space as it travels in all directions shown in Fig. 2.1. The idea of a world line was pioneered by Hermann Minkowski. Because the speed of light is invariant in postulate (ii), the light cone plays an important role in "causality". For a given event, all set of events that lie on or inside the past light cone can reach an observer. Events that do not lie within either a past or future light cone cannot be influenced in relativity.

The Lorentz transformation historically arose as a way to explain how the speed of light is independent of a given reference frame. For velocities of objects much less than that of the speed of light, the Lorentz transformation will reduce to a Galilean one. Special relativity, grounded in the formalism of the Lorentz transformation, helped to resolve paradoxes that arose in the nineteenth century with the discovery of electromagnetic phenomena where it was discovered that electromagnetic waves travel at a constant speed of light in vacuum. Individual laws for electromagnetism were effectively combined and captured in a famous set of equations now called the "Maxwell's equations".

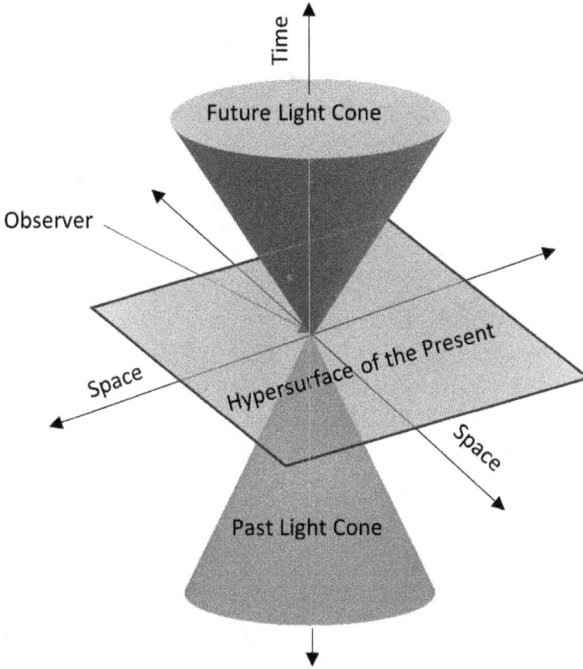

Figure 2.1. Schematic of a light cone in general and special relativity.

The introduction of a Lorentz transformation gives rise to the emergence of several physical implications from special relativity. A Lorentz formalism will in fact lead to such effects as time dilation (i.e., clocks will tick slower) and a length contraction (i.e., moving objects will shrink in size in the direction of movement) with observation of a moving object from a rest frame of reference. It also introduces the concept of relativistic energy for the moving particle. These results are somewhat counter-intuitive to normal experience on earth because of the lower velocities of objects that are typically much less than that of the speed of light.

Some results that are predicted by special relativity include the following phenomena:

(i) *Relativity of simultaneity*: Although two events may be simultaneous for one observer, they may not be simultaneous for another observer if the two observers are in relative motion.

(ii) *Time dilation*: Moving clocks are observed to tick more slowly than an observer's stationary clock.

(iii) *Length contraction*: Objects are observed to be shortened in their direction of motion with respect to a stationary observer.

(iv) *Maximum speed of things in nature*: No physical object, message or field line can travel faster than the speed of light in a vacuum.

(v) *Equivalence of mass and energy*: Mass and energy are equivalent. These two quantities are transmutable and proportional to the square of the speed of light.

(vi) *Composition of velocities*: Consider one frame at rest and another one moving at a velocity relative to the rest frame. A third frame is also moving but with a velocity relative to the second-moving frame. However, this third frame is not observed to move relative to the stationary frame with just a simple addition of the velocities as predicted by Newtonian physics. Instead, two Lorentz transformation must be applied resulting in a corrected velocity-composition law. If the two velocities are much less than the speed of light, then a simple addition law results.

Special relativity is a "falsifiable theory" so that it can be directly tested by experiment. The special theory has been verified in a number of tests. The Maxwell's equations for classical electromagnetic phenomena indicates that light moves with a characteristic velocity. This finding is an important underpinning in the postulate of item (iv) for special relativity. Originally, as discussed earlier, it was thought that light moved through a "luminous aether". The Michelson-Morley experiment failed to detect such an aether wind relative to the motion of the earth. FitzGerald and Lorentz subsequently proposed an ad hoc hypothesis with no theoretical basis. A "FitzGerald-Lorentz contraction" or "length contraction" was suggested to help explain the null result of the experiment. On the other hand, the Lorentz transformations, confirmed in the Michelson-Morley experiment, were derived from first principles by Einstein. A length contraction is a direct result of the predictions of special relativity in item (iii). In addition, the red-shift of light from a moving source was experimentally observed in 1938, and measured with better accuracy in 1941. This observation is in agreement with that predicted earlier by Einstein in 1905. These classic experiments have been repeated to higher degrees of accuracy over many years.

Other experiments have also provided confirmation of special relativity. Relativistic energy and momentum are seen to increase at high velocities in particle collisions in modern accelerators. Time dilation in item (ii) has been seen in a variety of experiments that were conducted in both the atmosphere and in particle accelerators. A relativistic "Doppler effect" occurs for observed frequencies with a red-shift with relative motion between a source and observer due to time dilation. This effect was confirmed by Ives–Stilwell in experiments in 1938. There have also been modern searches for "Lorentz violations". However, no deviations were seen from Lorentz invariance or symmetry. The laws of physics stay the same, in accordance with the fundamental "principle of relativity", for all observers that are moving with respect to one another within an inertial frame.

2.2 General Relativity

The general theory of relativity was completed in 1915. This theory was a generalization of special relativity. It established that accelerated motion and being at rest in a gravitational field are physically identical in accordance with the so-called "principle of equivalence". It recognized that free fall is simply inertial motion with no force acting on an object. This concept is in contrast to the classical theory of gravity proposed by Newton. In special relativity, inertially moving objects do not accelerate with respect to one other, whereas objects in free fall do accelerate. To resolve this apparent contradiction, Einstein proposed the concept of "curved space–time" in which he devised his general "field equations" that relate the geometry (i.e., the curvature) of space–time with the mass, energy and momentum of material within it. General relativity is a theory of gravity that involves the solution of the field equations. These equations are metric tensor constructs that capture the geometric and casual structure of space–time. The metric tensors define the topology of the space–time and how objects move through it.

Results predicted by general relativity include:

(i) *Precession of planetary orbits*: An orbit of a planet around the sun will precess differently from that predicted by Newton's theory of gravity. This effect is exemplified by the orbit of the planet Mercury that was used as a classic test of general relativity.

(ii) *Deflection of light in a gravitational field*: A ray of light will be bent as it passes near a large mass such as the sun. This result is also a classic test of general relativity.

(iii) *Gravitational red-shift of electromagnetic waves*: A ray of light travelling out of a gravitational well will lose energy. This energy loss corresponds to a decrease in the wave frequency and an increase in the wavelength toward the red end of the electromagnetic spectrum. This effect can be interpreted from the principle of equivalence, Doppler effect and gravitational time dilation. This is further considered a classic test of relativity.

(iv) *Shapiro time delay effect*: Radar signals passing near a massive object take slightly longer to travel to a target and return than if the mass had not been present. This result is usually considered a fourth classic test of relativity.

(v) *Gravitational time dilation*: In a gravitational field, clocks will run slower where stronger gravity will curve space–time so that time will proceed slower.

(vi) *Frame dragging*: A rotating mass can drag the space–time continuum along with it.

(vii) *Expansion of the universe*: General relativity can describe the expansion of the universe. This expansion causes distant galaxies to recede from us at a rate faster than the speed of light if the cosmological time and proper distance are used to calculate the speed of these galaxies. Cosmological time is used in the Big Bang model, where the expanding universe has the same density everywhere at a given moment of time. The proper distance is where a distant object would be at a specific moment of cosmological time. "Comoving coordinates" are a coordinate system used in cosmology theory that expand in the same way as space so that galaxies always remain at the same position.

As mentioned, general relativity has been confirmed by many classic tests. These classic tests include the precession of the perihelion for the orbit of the planet Mercury, the deflection of a light ray passing by the sun and the red-shift of a light ray in a gravitational field.

Before general relativity, there was a discrepancy between the classic prediction and the observed shift of the perihelion of Mercury by 43 seconds of arc per century. In fact, a planet inside of the orbit of Mercury (called Vulcan) was proposed to explain this observed difference. However, general relativity gave a theoretical prediction of 42.98 seconds of arc per century that accurately accounted for this discrepancy.

The first expedition to record the bending of light by the sun during a solar eclipse was made in 1919 by Sir Arthur Eddington. Since then there have been many measurements made with deflections ranging between

0.7 and 1.55 times the value predicted by Einstein. More accurate interferometric techniques have been used to measure the deflection of electromagnetic radiation from quasars passing near to the sun. Early measurements suggested deflections of 1.57 to 1.82 ± 0.2 arcseconds, supporting Einstein's prediction of 1.75 arcseconds. The gravitational bending of light can lead to a phenomenon known as "gravitational lensing". A gravitational lens results when a cluster of galaxies between a distant source and an observer bends the light from the source as it travels towards the observer. In 1979, this effect was confirmed by the observation of a twin quasar, namely Twin QSO SBS 0957+561. When the massive lensing object and the observer are aligned, the original light source will appear as a ring around the lensing object if the lens itself has a circular symmetry. If there is a misalignment, an arc segment will be seen instead.

Gravitational red-shifts from the sun and in terrestrial experiments exploiting the Mössbauer effect were verified in the 1960s. A gravitational red-shift was confirmed in 1976 with a hydrogen maser on a rocket launched to a height of 10,000 km as compared to an identical clock on the ground. Moreover, a solar red-shift was accurately measured using iron spectral lines in sunlight reflected from the moon in 2020.

In 1964, Irwin Shapiro proposed that a time delay should occur when radar pulses pass by the sun using the Schwarzschild solution of the Einstein field equations. Shapiro showed that the time delay for a radar signal travelling from the Earth to Venus and back would be about 200 microseconds. Tests performed in 1966 and 1967 using the MIT Haystack radar antenna match these predictions. Such experiments have been further repeated many times. A number of other tests have also confirmed the equivalence principle and frame dragging.

Chapter 3

Quantum Theory

"Quantum mechanics" in Section 3.1 describes the behaviour of particles. On the other hand, "quantum field theory", in Section 3.2, treats particles instead as excited states of their underlying quantum fields. Quantum field theory provides a seamless way to integrate quantum mechanics with special relativity. A coupling of quantum mechanics with general relativity is further described in Section 3.3 based on the use of a loop quantum gravity approach.

In quantum mechanics, there is no specific mechanism for creating and destroying particles, whereas, in quantum field theory, a field theory is used to model the transformation of such particles. Physical variables of position and momentum are found to be "operators" in quantum mechanics for transformations on the given quantum state. In quantum field theory, these quantities are just numbers. In essence, what is described as "a wave function" in quantum mechanics is replaced by a field that has the ability to create and annihilate particles. Although physical variables in quantum mechanics are quantized as discrete entities, it is the field itself and the "conjugate momentum field" that are quantized instead. As such, the particle number and type are not fixed in quantum field theory. Infinities have been observed to arise in calculated quantities of quantum field theory that can be cancelled or tamed through the use of a process called "renormalization".

3.1 Quantum Mechanics

Quantum mechanics is a fundamental theory of physics that describes the properties of matter at the atomic and subatomic scale. "Classical physics", which was developed beforehand, instead related the properties of matter to macroscopic phenomena. In quantum mechanics, quantities of a system, such as energy and momentum among others, are restricted to discrete values. In addition, objects can have the characteristics of both a particle and wave, or a so-called "wave-particle duality". There are restrictive limits on the accuracy to which a physical quantity can be predicted prior to measurement, given its initial conditions, in accordance with an "uncertainty principle". The "Pauli exclusion principle" also states that two or more identical particles with half-integer spins (such as electrons) cannot occupy the same quantum state of a given system. Quantum mechanics in effect is a trade-off in predictability between different measurable quantities as dictated by the uncertainty principle. No matter how careful measurements are made, one cannot precisely determine the measurement of a particle's position and momentum at the same time.

Quantum mechanics arose out of the solution of the black-body problem. The original theory of quantum mechanics was formulated in the 1920s. A more modern theory led to a mathematical formalism with the concept of a "wave function". This function relates information about a particle's energy, momentum and other properties with a probability amplitude that cannot be predicted with certainty. A rule proposed by Max Born gives a probability density function for the position of a particle when a measurement is made. The "Schrödinger equation" further relates the collection of amplitudes that pertain to one moment in time to that of another with the time evolution of the quantum state.

The "Dirac equation" proposed in 1928 is the relativistic version of the "Schrödinger equation" for electrons. It provided an important coupling and integration of special relativity with quantum mechanics. It also predicted the existence of "antimatter", with the existence of anti-electrons or positrons, discovered by Carl Anderson in 1932. The theory was validated with a proper prediction of the hydrogen spectrum. It also provided theoretical justification for the particle spin that is an intrinsic form of angular momentum carried by elementary particles.

Another consequence of quantum mechanics is interference phenomena with wave-particle duality. For instance, in the famous double-slit experiment, a coherent light source illuminates a plate containing two slits and produces an interference pattern of light on a screen behind the plate. Because of the wave nature of light, the interference pattern leads to bright and dark bands on the screen consistent with the wave nature of light. This pattern is akin to the passage of a water wave through nearby openings, giving rise to constructive and destructive interference patterns with waves of different heights. On the other hand, if one monitors which slit a photon passes through, then a different pattern results consistent with classical results.

In addition, as observed in the radioactive decay of an atom, a particle can tunnel through an energy barrier even if its kinetic energy is less than the maximum value of the potential energy barrier of the nucleus. Such phenomena is seen, for instance, during alpha decay with a tunnelling of alpha particles through the Coulomb barrier of the nucleus [Lewis *et al.*, 2017].

In addition, a feature of quantum mechanics that differs from classical theory is the consequence of "quantum entanglement". Here a whole system becomes intertwined so that the individual parts of a system cannot be distinguished or described independently of the state of the others. This effect can occur even if particles are separated at very large distances.

Quantum mechanics requires an advanced mathematical formulation to describe these phenomena. It involves the description of a vector space that requires the combined mathematical methods of linear algebra, complex numbers and calculus.

3.2 Quantum Field Theory

Quantum field theory is an amalgamation of the concepts of classical field theory, special relativity and quantum mechanics. It provides for an understanding of subatomic particle physics and condensed matter physics. In quantum field theory, particles are importantly considered to be just excited states of their underlying quantum fields. The particle interactions, themselves, involve interactions among the different particle fields.

Quantum field theory grew out of the need to describe interactions between light and electrons. This work culminated in a relativistic quantum field theory of electrodynamics called "quantum electrodynamics" developed in the 1920s. It thereby accounted for how electrically-charged particles interact by means of an exchange of photons. The appearance of infinities in the 1950s in the early calculations necessitated the need for the development of a renormalization process to eliminate these infinities. The application of so-called "gauge theory", where the dynamics of a system do not change under local transformations that preserve the symmetry of the physical system itself, provided a way to encompass both weak and strong nuclear interactions. This development led to the "Standard Model" of particle physics in the 1970s.

The theory of classical electromagnetism culminated with the completion of Maxwell's equations in 1864. This set of equations accounted for the relationship between the electric and magnetic fields, the electric charge and current. It showed that the electric and magnetic fields propagate at the speed of light. Despite its success, it failed to account for the observed discrete lines in atomic spectra or for the spectral intensity of black body radiation. Building on Max Planck's idea of a "quantum harmonic oscillator" concept for discrete energy values, Albert Einstein proposed in 1905 that light is composed of individual packets of energy called "photons". This idea provided an explanation for the photoelectric effect. Niels Bohr in 1913 further showed that electrons within atoms have discrete energy levels. These ideas were then incorporated into a more coherent theory of quantum mechanics.

With the use of a wave function, the Schrödinger equation was further proposed to calculate the probability amplitude for the possible outcomes of measurements made on a system. This equation was not consistent with special relativity and could not generally explain the spontaneous emission of photons. It treated spatial variables as linear operators but time was just a number. The linear operator itself is a generalized and physically-based construct used in quantum mechanics for turning a function into another one rather than a number. With the further work of Max Born, Pascual Jordan and Werner Heisenberg, a quantum theory in 1925–1926 was developed which was able to treat the electromagnetic field without any interaction with matter. In this theory, the electromagnetic field acted as quantum harmonic oscillators. A year later, quantum electrodynamics was introduced with the addition of an interaction term to the theory to explain the spontaneous emission of electrons. Here, quantum fluctuations, which occur even in a perfect vacuum in accordance with the uncertainty principle, can stimulate such emission. This theory was therefore able to account for the scattering of photons, resonance fluorescence and non-relativistic Compton scattering. In 1928, Dirac further presented a wave equation for relativistic conditions. This equation permitted the existence of unphysical negative energy states. However, by 1929, it was realized that such unreal states could be removed with the existence of particles of the same mass as electrons but with opposite charge. This theory gave rise to the prediction of antimatter with the eventual discovery of positrons in 1932. The Dirac theory also explained the 1/2-spin of the electron. It also permitted a calculation of the fine structure of the hydrogen atom as well as yielded a relativistic equation for Compton scattering.

Quantum electrodynamic calculations were plagued by resultant infinite quantities. This shortcoming arose in perturbative calculations using the theory, including the electron self-energy and the vacuum zero point energy of the electron and photon fields. In 1950, it was suggested by Julian Schwinger, Richard Feynman, Freeman Dyson and Shinichiro Tomonaga to replace the infinite calculated values, i.e., the mass and charge of the electron, by their finite values in a process called "renormalization". This approach was then able to correctly predict the anomalous magnetic moment of the electron. At the same time, Feynman introduced the concept of the "path integral formulation". A visual methodology was developed using "Feynman diagrams" to compute terms in the perturbative expansion calculation of the scattering amplitude for particles undergoing interaction.

A set of rules was developed by Feynman for calculating the quantum amplitude associated with these diagrams. For example, for a line representing a photon this gives rise to a "photon propagator". Similarly, a line representing an electron yields an electron propagator and so on. The point where a photon line joins an electron line is called an "interaction vertex", where a coupling factor for the electron charge e is applied. Thus, following the detailed Feynman rules (for example, see page 534 in [Zee, 2010]) and multiplying the various factors together, quantum amplitudes can be evaluated.

The renormalization process, however, had a limited applicability. In quantum electrodynamics, the infinities in the perturbative calculations could be eliminated by redefining a small number of physical quantities such as the mass and charge of the electron. This approach is only possible though with quantum electrodynamics, where other theories such as that proposed by Enrico Fermi for weak interaction processes were found to be "non-renormalizable". For the Feynman diagrams, a perturbative calculation can be performed with a requirement that there is a small number for the "coupling constant" in the series expansion. For instance, in quantum electrodynamics, the coupling constant is the fine structure constant with a value of $1/137$. On the other hand, in strong interactions, this constant is of the order of unity so that reliable calculations are not possible in quantum field theory.

In 1954, the local symmetry of quantum electrodynamics was generalized using a so-called "Yang–Mills" (non-Albelian) gauge theory. This theory used more complicated symmetry groups involving the special unitary (SU) group theory to describe the behaviour/symmetry of elementary particles. This approach includes the unification of the electromagnetic force and weak force (i.e., U(1) × SU(2) group theory) as well as the strong force (i.e., SU(3) group theory). This combined approach, in fact, led to the fundamental basis of the "Standard Model" of particle physics in Chapter 4.

Quantum field theory has made highly accurate predictions in agreement with experiment. For instance, this theory has predicted the "intrinsic magnetic moment" of the electron that describes how a fast spinning electron precesses in a magnetic field. The ratio of the magnetic moment to its angular momentum is known as the "g factor". As of 2011, the g factor has been measured to a great accuracy of one part in a trillion, i.e., $g_e/2 = 1.001\ 159\ 652\ 180\ 73(28)$. In comparison, the prediction from quantum field theory agrees with the measured value to 12 significant digits — making it one of the most accurately verified predictions in physics!

3.3 Loop Quantum Gravity

Loop quantum gravity combines quantum mechanics and general relativity to provide a consistent theory of "quantum gravity". It allows a means to incorporate matter in the Standard Model into a framework based on Einstein's geometric formalism rather than considering gravity as just a force. As such, this theory postulates that the structure of space and time is composed of finite loops that are woven as a fine structure of "spin networks". The evolution of the spin network, or so-called "spin foam", has a scale at the order of the Planck length of 10^{-35} m. The "spin connection" is a quantum operator. It describes the dynamics of space–time and ensures that the quantum states of the system are consistent with classical general relativity. This operator therefore guarantees that the behaviour is determined by underlying quantum mechanical and geometric properties.

Loop quantum gravity considers space–time as a dynamical field that is a quantum object. Here the quantum discreteness affects the structure of space–time itself. As such, space has a granular structure on the order of the Planck length. The concept of a spin network was originally introduced by Roger Penrose as a mathematical abstract structure. However, as shown by Carlo Rovelli and Lee Smolin in 1994, the quantum operators of the theory associated to area and volume have a discrete spectrum so that the geometry is quantized. Thus, Rovelli and Smolin were able to define a non-perturbative and background-independent quantum theory of gravity. This "background independent theory" does not depend on space and time, except for its invariant topology. It is independent of the coordinate system (i.e., a so-called "diffeomorphism invariance") on the shape of space–time and the values of the various fields in space–time. This work therefore led to loop solutions, where the spin network is a direct representation of the quantum state of space–time that evolves over time in discrete steps.

The theory is based on the reformulation in 1986 of general relativity using "Ashtekar variables". This representation recognizes geometric gravity as a mathematical analogue of electric and magnetic fields in accordance with "Yang–Mills" theory as part of the Standard Model of particle physics. Several versions have been proposed to describe the dynamics of the theory. In the more traditional version, one considers a canonical quantization of general relativity using the "Wheeler–Dewitt equation" that is an analogue of the Schrödinger equation. In the newer covariant spin foam approach, developed over several decades and completed in 2008, the quantum dynamics involve a sum over discrete versions of space–time.

The most well-developed application of loop quantum gravity is "loop quantum cosmology". This latter theory provides a means to study the early universe. It is also successful in explaining some key features of quantum gravity, including the quantization of area and volume and the absence of space–time singularities. However, it still faces challenges as a fully covariant and completely background-independent formulation. It also lacks an experimentally-verifiable prediction of quantum gravity effects.

Chapter 4

The Standard Model

The Standard Model is based on quantum field theory and predicts almost all known particles and forces in nature other than gravity with high accuracy. It provides a model of particle physics with a theoretical framework for describing three of the four forces of nature (Section 4.1) along with a classification of elementary particles (Section 4.2). The model was developed in the latter half of the 20th century. Its current formulation was presented in the mid 1970s with experimental confirmation of quarks. This confirmation included the most recent and important discoveries of the top quark (1995), the tau neutrino (2000) and the Higgs boson (2012).

4.1 Fundamental Forces of Nature

This model describes three of the four fundamental forces of nature, including:

(i) Electromagnetic interactions,
(ii) Weak interactions,
(iii) Strong (nuclear) interactions.

A "boson" is a subatomic particle that is indistinguishable from one another (i.e., follows "Bose–Einstein" statistics) whose spin (angular momentum) has an integer value $(0, 1, 2, \ldots)$. A force-carrying particle, called a "gauge boson" mediates the electromagnetic, weak and strong forces as spin-1 particles. The forces in nature result from an exchange of gauge bosons, which are the quanta of a given field. The number of gauge bosons depend on the symmetry group of the field. The symmetry groups for particle physics belong to a so-called "unitary group" and "special unitary group", which give rise to a specific number of "generators" for the field.

The number of generators for the unitary group U(1) is equal to one. For the special unitary group SU(N), it is the square of the number N shown minus one.

4.1.1 *Electromagnetic force*

The symmetry group for the electromagnetic force is called U(1). Since it has a single generator, the force is mediated by a single massless particle known as a photon that has spin-1 with two polarization states.

4.1.2 *Weak force*

The gauge group for the weak force is the special unitary group SU(2). Thus, there are $2^2 - 1 = 3$ physical gauge bosons that mediate the weak force. These spin-1 particles are: W^+, W^- and Z^0 that have three polarization states. The mass of a particle can be equivalently related to its measured energy in accelerator experiments in accordance with Einstein's energy–mass relation in Chapter 2 by dividing the energy of the particle by the speed of light squared (c^2). Here the energy of the particle can be measured in the units of Giga-electron volts (i.e., GeV or 10^9 eV). The W^+ has a mass of 80 GeV/c^2 and carries a plus 1 charge. The W^- has a mass of 80 GeV/c^2 and carries a minus 1 charge. The Z^0 has a slightly larger mass of 91 GeV/c^2 and is electrically neutral.

4.1.3 *Strong force*

The gauge group for the strong force is special unitary SU(3), yielding $3 \times 3 - 1 = 8$ gauge bosons called "gluons". These massless spin-1 particles have two polarization states. Gluons mediate the charge of the strong force called "colour", which, for instance, is responsible to bind neutrons and protons together in the nucleus of an atom. The theory that describes the strong force is called "quantum chromodynamics".

4.2 Particle Physics

In the early development of the quantum theory of electrodynamics, it was shown that electrically-charged particles interact via an exchange of photons (which are electrically-neutral). The unitary U(1) transformation for the electromagnetic force is in fact commutative thereby reflecting the symmetry for rotation around a circle. For instance, if one rotates by a certain angle around a circle, and then rotates from this point on by a second rotation, the combined effect is equivalent to rotating by the second angle first and then by the first one. As such, this type of symmetry is commutative (i.e., Albelian). However, for other types of particle interactions, different symmetry groups involving "non-Albelian gauge transformations" are required. Contrary to the commutative U(1) group for electrodynamics, the elements in the non-Albelian groups do not obey a commutative law. As such, particles in other symmetry groups for the other forces carry a new type of "charge". Here, the particles interact with the exchange of *massless* (force-carrying) "gauge bosons" in contrast to that of a photon. In 1960, Sheldon Glashow developed a non-Albelian gauge theory, which was able to successfully unify both the electromagnetic and weak interactions. This same result was obtained independently in 1964 by Abdus Salam and John Clive Ward. Through a mechanism of "spontaneous symmetry breaking", the original massless gauge bosons are able to importantly acquire a mass, which currently explains the experimental observations. Steven Weinberg then in 1967 combined these various ideas into an "electroweak theory", which combined electromagnetism and the weak interaction. This theory was now able to account for interactions among all leptons (i.e., electrons, muons and neutrinos) and the so-called "Higgs boson". The Higgs boson is in fact a quantum excitation of the so-called "Higgs field". Furthermore, in 1971, it was discovered that some strong interactions could also be explained again with a non-Albelian gauge theory, giving rise to the theory of "Quantum chromodynamics". This theory, however, is only specific to high-energy particle interactions, where one needs a small value of a "coupling constant" in the theory. At higher energies, the coupling constant decreases. A smaller coupling constant specifically allows for perturbative methods to be used for interaction calculations for validation of the theory.

The full theory, involving the electroweak theory and quantum chromodynamics, is referred today as the "Standard Model" of elementary particle physics. This theory is therefore able to explain all fundamental interactions except for the force of gravity. The theory has yielded remarkable predictions with experiment. In this model, there are two basic types of particles:

(i) *Bosons*: Here spin-1 gauge bosons are the particles responsible for transmitting the electromagnetic, weak and strong force. These particles include the photon that transmits the electromagnetic force, the $W^\pm$ and Z particles that carry the weak force, and gluons that account for strong nuclear particle interactions.

(ii) *Fermions*: A fermion is a particle that follows "Fermi–Dirac" statistics having a half-odd-integral spin $(1/2, 3/2, 5/2, \ldots)$. These particles importantly obey the Pauli exclusion principle. Matter is composed of spin-1/2 fermions such as electrons.

The standard model also requires the introduction of a spin-0 boson called the "Higgs boson". As mentioned, particles that interact with the associated Higgs field acquire mass through a mechanism of spontaneous symmetry breaking of the electroweak field (see Fig. 4.1). The Higgs field has a non-zero strength everywhere even in an otherwise empty space.

All constituent components of the theory have now been discovered in particle accelerator experiments. The standard model of particle physics is shown in Fig. 4.2. It includes 12 force carriers (which are bosons): the eight gluons, W^+, W^- and Z^0, and the photon. There are also many matter particles (which are fermions). The matter particles are of two types: (i) 12 leptons (which include the electron e^-, muon μ^-, τ^-, and associated neutrinos: ν_e, ν_μ, ν_τ and their antiparticles; and (ii) 36 quarks which come in "flavors" of: up (u), down (d), charm (c), strange (s), top (t) and bottom (b), where each of the six quark flavors come in three colours (red, green and blue), for a total of 18 particles × 2 for each of their antiparticles. The proton, neutron and many other elementary

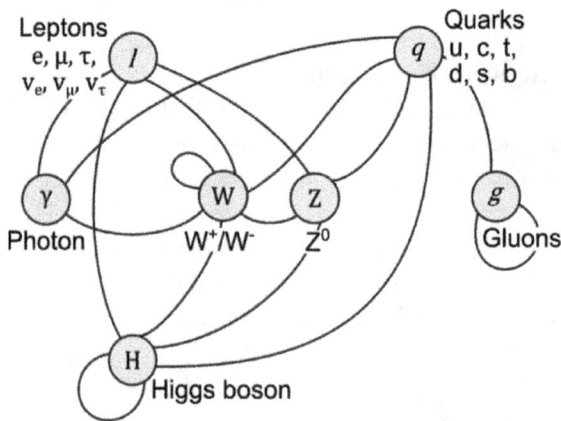

Figure 4.1. Summary of interactions between particles in the standard model.

```
                          ┌─────────────────────────┐
                          │   Elementary particles   │
                          └─────────────────────────┘
```

Elementary fermions	Elementary bosons
Half-integer spin	Integral spin
Obey Fermi Dirac statistics	Obey Bose-Einstein statistics

Quarks & antiquarks	Leptons & antileptons	Gauge bosons	Scalar bosons
Spin = ½	Spin = ½	Spin = 1	Spin = 0
Colour charge	No colour charge	Force carriers	
Strong interactions	Electroweak interactions		

Three generations	Three generations	Exchange particles for force interactions	Unique
1.Up (u) (+2/3), Down (d) (-1/3)	1.Electron (e^-) (-1), Electron neutrino (v_e) (0)	1.Photon (γ) (0); electromagnetic interaction	Higgs boson (H^0) (0)
2.Charm (c) (+2/3), Strange (s) (-1/3)	2.Muon (μ^-) (-1), Muon neutrino (v_μ) (0)	2.W and Z bosons (W^+ (+1), W^- (-1), Z (0)); weak interaction	
3.Top (t) (+2/3), Bottom (b) (-1/3)	3.Tau (τ^-) (-1), Tau neutrino (v_τ) (0)	3.8 types of gluons (g) (0); strong interaction	

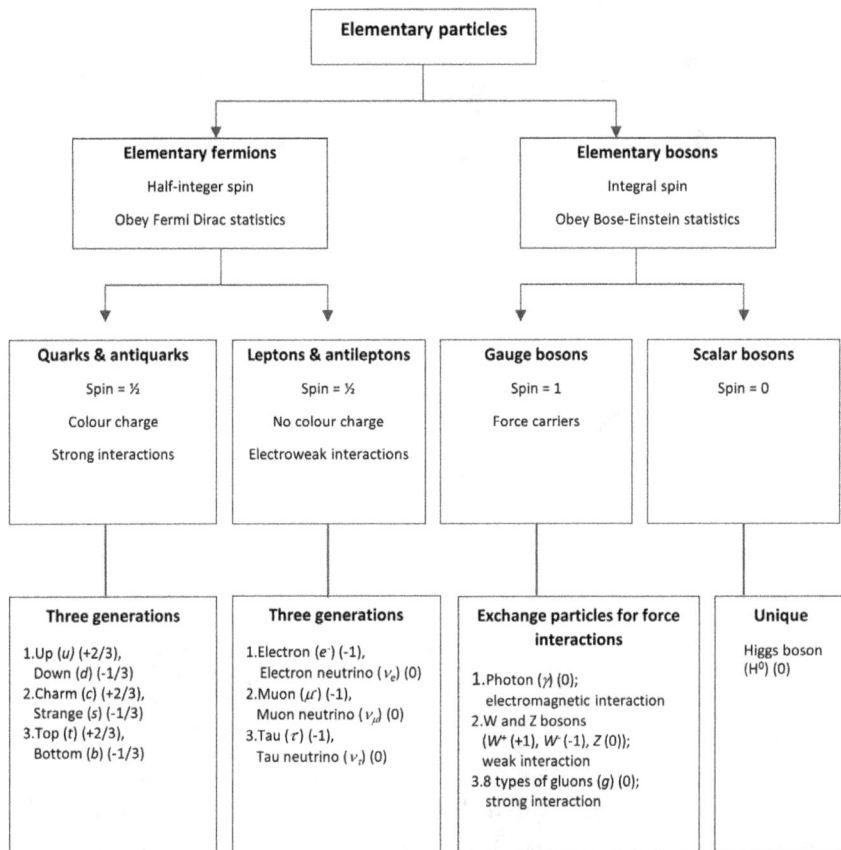

Figure 4.2. Elementary particle model. Note that electrical charge in units of e is given in brackets.

particles of matter are made of quarks (see Fig. 4.2). Only combinations with no net colour can exist as free particles (see Fig. 4.3). A quark and its anti-quark form an unstable particle called a "meson". Three quarks, one of each colour, form stable particles called "baryons" of which protons and neutrons are examples. The concept of a colour charge was needed in the quark model to ensure consistency of the Pauli exclusion principle. Adding the leptons and quarks together give 48 matter particles. Finally, adding the matter particles and force carriers together give a total of 60 particles in the Standard Model.

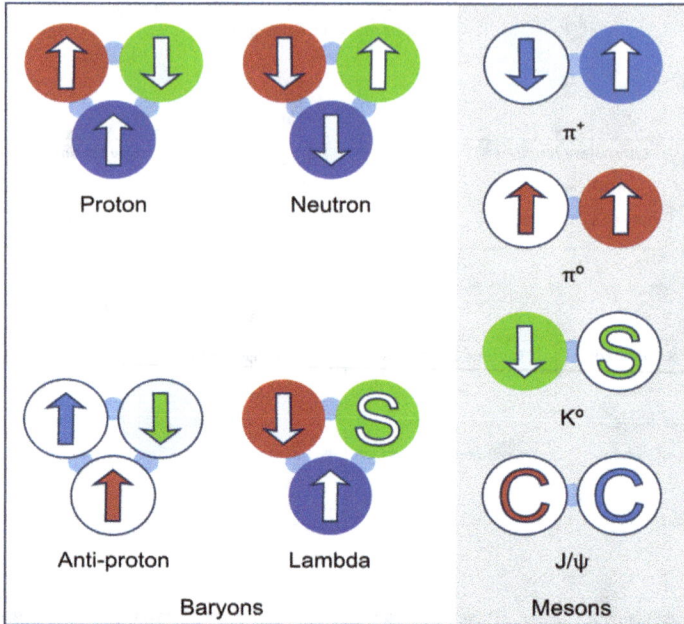

Figure 4.3. Baryons and mesons in the quark model.

As mentioned, quarks are elementary particles that combine in groups of twos and threes to form hadrons. However, more recently in 2022, observations were announced in the Large Hadron Collider experiment at the CERN European Organization for Nuclear Research that never-before-seen "exotic particles" had been detected. These particles included a new kind of "pentaquark" and the first ever duo of "tetraquarks" as five- and four-quark particles.

As shown in Figs. 4.1 and 4.2, there are three known types (flavors) of neutrinos [electron (ν_e), muon (ν_μ) and tau (ν_τ)]. It has been recently confirmed that neutrinos can oscillate and switch identities from one flavor to another in atmospheric experiments with the Super-Kamiokande detector in Japan and experiments at the Sudbury Neutrino Observatory in Canada for neutrinos coming from the sun. This work was awarded the recent 2015 Noble Prize in Physics to Arthur MacDonald and Takaaki Kajita. This result dispelled the long held notion that neutrinos are massless, thereby challenging the Standard Model of particle physics. Neutrinos are the second most abundant particle in the universe after photons and thus have an important effect on the evolution of the universe and how much mass is in the universe.

Recent evidence suggests that about two thirds of the mass of the universe consists of "dark energy" that is causing its accelerated expansion. "Dark matter" affects the gravitational pull of clusters of galaxies. Baryonic and non-baryonic (dark) constituents make up the rest of the matter in the cosmos.

Chapter 5

String Theory

Currently a single "unified field theory" that encompasses gravity does not exist despite the fact that quantum physics and general relativity were significant theoretical developments in the twentieth century. Quantum calculations are particularly plagued by the occurrence of infinities. This result happens because particle interactions take place at a single mathematical point of zero distance. On the other hand, string theory may offer greater possibilities where interactions do not occur among point particles. Hence, this more generalized theory may possibly offer a unified theory of both particle physics and quantum gravity.

The historical development and theoretical framework of string theory, including higher-dimensional objects, are given in Section 5.1. The classification of five different types of open/closed strings with different particle statistics and orientations, dualities that interlink some of these classifications, as well as their unification in M theory, is detailed in Section 5.2. The possible application of string theory to various disciplines of theoretical physics, as well as recent experimental investigations into its validity, is further discussed in Section 5.3.

5.1 Fundamentals of String Theory

An early approach for such a model was a classical theory that unified both gravity and electromagnetism with the idea of an extra fifth dimension. In 1921, Theodor Kaluza published an extension of the theory of general relativity from four to five dimensions. The metric tensor used in general relativity was extended to 15 components, consisting of 10 components for four-dimensional space–time, 4 components for electromagnetism, and one component for a so-called "dilaton" field. These equations were therefore able to yield the Einstein field equations, the Maxwell equations for the electromagnetic field, and an equation for a scalar field. Furthermore, as an advancement of the theory, Oskar Klein introduced in 1926 the concept of a fifth dimension that was curled up as a circle with a very tiny radius of 10^{-30} cm.

However, the Kaluza–Klein theory is incomplete since it cannot be generalized to include the weak and strong forces nor provide a quantum theory of gravity. However, this theory with an extra dimension is an interesting idea although there was no understanding of quantum field theory back in the 1920s. The Kaluza–Klein approach may in fact be thought of as an early precursor for a unification of fundamental forces. This framework of employing extra dimensions has re-emerged in more recent times as "String theory" as a possible candidate for a unified field theory.

In string theory, fundamental particles are not points but rather strings, which are extended one-dimensional objects. The strings can be either open objects or closed upon themselves with the ends connected as a loop as shown in Fig. 5.1. The strings can propagate through space–time and interact with one another. The excitations (i.e., vibrations) of the string produce the fundamental particles. Since these objects have a finite length on the order of the so-called "Planck scale" of 10^{-33} cm, when viewed at distances larger than the string scale, a string looks like any ordinary particle. Its mass and charge among other properties are determined by the vibrational states of the string.

String theory is also a theory of "quantum gravity" that can describe gravity at very small distances. One of the many vibrational states of the string leads to the presence of a "graviton". The graviton is a quantum

Figure 5.1. Schematic of an open and closed string in string theory.

mechanical particle that carries the gravitational force. However, string theory lacks experimental proof as a "falsifiable theory". Critics of the theory have argued that it cannot make predictions that are experimentally testable, such as in general relativity, where classic tests of general relativity have included the bending of light, the gravitational red-shift and the advance of the perihelion of Mercury. On the other hand, some scientists believe that it has the underpinnings for a "theory of everything". A fully complete theory would encompass all of the four fundamental forces of nature, as well as all forms of matter, with a unified description of gravity and particle physics under a single unified mathematical structure.

The vibrating string is a single underlying entity that has different modes that can give rise to a plethora of particles. Each mode appears as a different particle, where one such mode could be an electron while another may be a quark. Hence, the string can split apart or combine with others providing a simple conceptual framework. For example, if one string splits into two, the resulting daughter strings may vibrate at different modes corresponding to different particles thereby representing the process of radioactive decay. On the other hand, strings may join up, which would be consistent with particle absorption.

String theory relies on the existence of extra dimensions. These extra dimensions may help explain the classification of different families of particles. For instance, for the lepton family of particles, there are electrons along with two other heavier particles known as the muon and tau particles, along with their corresponding neutrinos as shown in Fig. 4.1. The same situation exists for quarks with again three families of particles in Fig. 4.1. It is also not clear how different particle interactions occur. However, the higher spatial dimensions in string theory may provide a physical answer to such quandaries. The numbers and types of particles seen in the universe may physically depend on how extra dimensions are compactified in the topology of the universe. With "compactification", extra dimensions are assumed to curl up in order to become very small so that one obtains a theory that effectively has a lower number of dimensions. This process may relate to how the strings wrap around the compactified dimensions as to what vibrational modes and particles are possible. An important manifold in topology theory is the Calabi–Yau manifold. This manifold compactifies six spatial dimensions leaving three macroscopic dimensions plus time. This configuration would give rise to a 10-dimensional universe in string theory. Moreover, a key feature of the Calabi–Yau manifold is that spontaneous symmetry breaking can also occur. This feature is

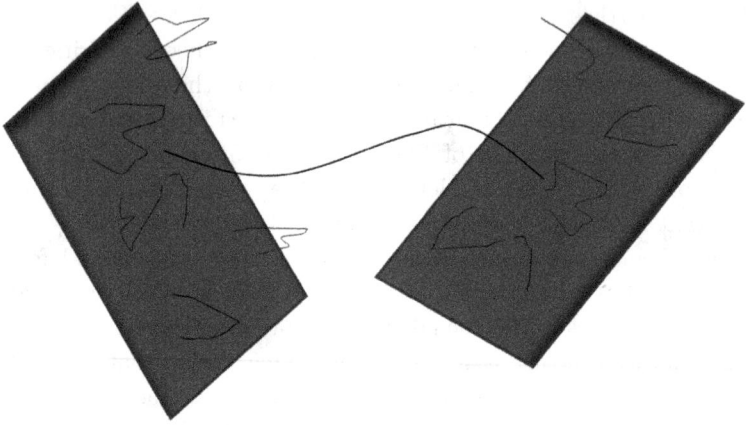

Figure 5.2. Schematic of open strings attached to a pair of *D*-branes.

an important phenomenon needed in quantum field theory and the standard model of particle physics as described in Chapters 3 and 4.

In addition, the loose ends of strings can attach to higher-dimensional objects called "*D*-Branes" as shown in Fig. 5.2. The symbol *D* refers to the dimension of the brane. It also captures generally the mathematical condition for the end points of the string that lie on a brane that are fixed as a "Dirichlet boundary condition". The word "brane" itself stems from the notion of a (two-dimensional) membrane. A string can be thought of as a one-dimensional brane (1-brane). A 2-brane is simply a higher-dimensional membrane surface like that seen on a drum. Even higher-dimensional objects exist as a *p*-brane with *p*-spatial dimensions. Thus, the brane structures provide an object that the fundamental strings can attach to.

Quantum fields, such as the Yang–Mills theory of electromagnetism, involve strings that are attached to D-branes. This concept is key because it is in contrast to gravitons that are not attached to D-branes. The gravitons being quanta of gravity can travel and easily *leak off* of a D-brane. This formalism may help explain and conceptualize why electromagnetism (among the other forces) are so much stronger than gravity. Thus, the universe can be thought of as having a three-dimensional brane embedded in a higher dimensional space–time "bulk". Physical manifestations of the electromagnetic force, among the other two forces, are strings stuck to a brane. The world is also experienced as having three-spatial dimensions.

Gravity, on the other hand, is mediated by strings that leave the brane and travel off into the bulk thereby producing a weaker observed force.

5.2 Classification of Strings and Dualities

Five different types of strings have been proposed. However, it has been further shown that these approaches are simply different ways of looking at the same theory through the concepts of "M theory" and "dualities". The five different types of strings originally included:

(i) **Bosonic string theory:** This simplest formulation of string theory only pertains to boson-type particles. Since this theory avoids fermions, it is not complete as it does not describe matter in the universe. This theory includes both open and closed strings that can be *oriented* or *unoriented*. An oriented string indicates that there is a specific direction that the vibration modes can travel on the string. The bosonic string state also includes a "dilaton" state as a gravitational scalar field with a spin-2 graviton and a coupling constant for determining the interaction strength of the field. Interestingly, the dilaton may play a role in the non-zero "cosmological constant", introduced in Einstein's field equations accounting for the expansion of the universe. However, in order to avoid so-called non-physical "ghost" states (i.e., states with a negative norm) in the theoretical framework, the theory requires 26 space–time dimensions for mathematical consistency. The ground state of the boson string itself (or lowest excitation mode of the string) has a negative squared mass. As such, the vacuum is an unstable "tachyon" particle. A tachyon particle in relativity can travel faster than the speed of light. Since, in this simple theory, there is no way to eliminate tachyons, one is forced to consider superstring theories.

(ii) **Type I string theory:** This theory is more general and considers both boson- and fermion-type particles. This particular theory relies on the idea of "supersymmetry". In this concept, the equations for the force and the equations for matter are identical. With supersymmetry for the two classes of particles (i.e., bosons and fermions), each particle from one class has an associated "superpartner" in the other class. Consequently, this theory has a gauge group symmetry of $SO(32)$ so that the number of dimensions is reduced to 10 space–time dimensions for mathematical consistency in order to eliminate the ghost states.

It includes both open and closed strings that are unoriented. However, no super-partner has ever been experimentally seen to date. Hence, either supersymmetry does not exist in nature, or it has been broken where the superpartners are so massive that particle accelerators are presently unable to reach such high energies.

(iii) **Type II A string theory:** This version of the theory also includes supersymmetry with open and closed strings. However, in this version, the open type II A strings have their loose ends attached to "*D*-Branes". The fermions in this theory are oriented. Since this theory only has a U(1) symmetry, it can only describe the phenomena of electromagnetism and gravity but not the weak and strong forces.

(iv) **Type II B string theory:** This type of string is similar to the type II A strings but have chiral fermions. However, it does not have a gauge symmetry and as such it cannot provide a unified theory of physics.

(v) **Heterotic string theory:** This version includes supersymmetry but only allows for a closed string (or loop) as a hybrid of a superstring and a bosonic string. There are two kinds of heterotic strings possessing SO(32) and $E_8 \times E_8$ group symmetries. Here the "left-moving" and "right-moving" excitations are decoupled since it is not possible to define boundary conditions with left-moving and right-moving excitations given their different characteristics. As such, left-moving excitations are treated as a bosonic string propagating in 26 dimensions while the right-moving excitations are superstrings in 10 dimensions.

Thus, five different versions of string theory were originally developed. However, in the mid-1990s, it was realized by Edward Witten that all of these versions were different limiting cases belonging to a single theory with 11 dimensions known as "M theory".

The different string theories are further connected through "dualities". In essence, a duality refers to a situation where two seemingly different physical systems turn out to be equivalent in a non-trivial way. Two important dualities in string theory are the "*S duality*" and "*T duality*". A *T-duality* relates a theory with a small compact dimension to one where the same dimension is large. In fact, this duality relates type II A and II B theory, and also separately relates the two heterotic theories. Consequently, each of these types of groups are the same. In essence, as one transforms from a small to a large distance scale, one is exchanging momentum and winding modes and vice-versa. The winding modes describe how a string is wound around a compactified dimension. In addition, a second important duality is *S-duality*, which manifests itself through a coupling constant. This constant describes the strength of an interaction, which can be either a strong or weak force in string theory. By interchanging the sign of a dilaton field, one can change a large coupling constant into a small one. Thus, under a *S-duality*, a type I superstring is related to the SO(32) heterotic superstring theory. This duality means that a strong interaction in type I superstring theory is the same as a weak interaction in heterotic SO(32) theory, and vice versa. Thus, the two theories are equivalent but with different coupling strengths.

5.3 Application and Experimental Testing

String theory has found various applications. Importantly, it can be used to construct a variety of models for particle physics. Starting with 10- or 11-dimensional space–time or M theory, one can postulate a shape for the extra dimensions. Different shapes can account roughly for the standard model of particle physics. String theory has also been used for investigating the theoretical properties of black holes. A black hole, as discussed in Chapter 7, is a region of space–time where the gravitational field is so strong that no particle or radiation can escape from it. Such objects can arise when massive stars in galaxies undergo gravitational collapse. String theory has provided a general framework from which the thermodynamics of a black hole can be studied. The Bekenstein–Hawking formula for predicting the so-called "entropy" of a black hole, in accordance with classical thermodynamics theory, can be derived from the properties of D-branes. In this application, D-branes can be considered as fluctuating membranes when they are weakly interacting but can become dense massive objects when their interactions are strong. This property is indistinguishable from black holes. Furthermore, in late 1997, an additional relationship was discovered called the "anti-de Sitter/conformal field theory correspondence" (AdS/CFT correspondence). This correspondence was able to relate string theory to the quantum field theory of Chapter 3. An anti-de Sitter space is thought to have occurred in the very early stages of the universe (see Chapter 6). At this time, space–time had a negative curvature as a solution of Einstein's field equations for general relativity. It was discovered that the boundary of anti-de Sitter space can in fact be regarded as the required space–time for quantum field theory.

String theory, however, is somewhat controversial. It is not falsifiable, where the concept of extra dimensions has no experimental foundation at the extremely high energies needed in accelerator experiments to see such effects. String theory in fact does not contain the Standard Model in its lower-energy limit but consists of a higher-symmetric group that describes a system of particles and fields. This higher-symmetric group encompasses a "$N = 4$ supersymmetric Yang–Mills gauge theory". This gauge theory enables the dynamics of a system under a set of symmetries where the Lagrangian is invariant under local transformations. Thus, string theory requires a higher degree of symmetry despite the fact that no new partner particles as suggested in supersymmetry theory, with masses of up 1 TeV/c^2, have yet been discovered in experiments to date in the Large Hadron Collider.

Some criticisms of the theory recognize that these models require a shape for the extra dimensions of space–time. Each of these different shapes corresponds to a different universe (or vacuum state) for the large collection of particles and forces in the theory. Current string theory in fact has an enormous number of vacuum states. These states may be so diverse and great in number that they can accommodate any phenomena that may be observed at low energies. Also, it is not clear if string theory is completely compatible with an observed positive cosmological constant for the universe.

Thus, there is a lack of experimental evidence where the theory makes few testable predictions. The extra dimensions and large number of solutions again make it difficult to test the theory. Some solutions require the existence of very specific conditions where such fine-tuning makes it difficult to understand how such conditions could be realized in nature. Moreover, a large number of solutions make it unfalsifiable. Therefore, some argue that such criticisms for experimental evidence and physical uncertainty in the theory may therefore question string theory as a quantifiable unified theory for the four fundamental forces of nature.

Chapter 6

Cosmology

Cosmology involves the investigation of the large-scale structure, dynamics and fate of the universe. It depends on the disciplines of astronomy and particle physics. This field of study accounts for the presence of dark matter and energy, which is parametrized in the model of "Big Bang theory" known as the "Lambda (i.e., for inclusion of the cosmological constant)-cold dark matter (CDM) model". The universe is thought to have begun with the Big Bang about 13.8 billion years ago. It underwent an immediate expansion of space as a result of cosmic "inflation".

The history of cosmology is outlined in Section 6.1. The observational evidence from astrophysical studies is described in Section 6.2. The fundamental aspects of the theory and parameters of the model as they relate to the observable universe, which describe its expansion and evolution, is given in Section 6.3.

6.1 Historical

Modern scientific cosmology began in 1917 with Einstein's publication "Cosmological Considerations of the General Theory of Relativity" in his theory of general relativity. Willem de Sitter, Karl Schwarzschild, and Arthur Eddington investigated the astronomical implications of this theory. These works led to a realization that the universe is not static. The idea of an expanding universe, in fact, was suggested in 1922 by Alexander Friedmann.

A cosmological constant was originally introduced by Einstein in his theory of general relativity to provide a static model of the universe. However, it was realized that the universe is in fact expanding. This led to the introduction of the Big Bang model in 1927 by Georges Lemaître. Historically, Copernicus rejected the anthropocentric view that the Earth is at the centre of the universe. Instead, he postulated that the Sun is at the centre, with the planets orbiting around it. From Newton's law of gravity, Kepler realized that the planets move on elliptical orbits.

A Cepheid variable star is a class of star that pulsates with changes in its brightness exhibiting a well-defined stable period and amplitude. A strong relationship exists between the variable luminosity of a Cepheid star and its pulsation period. In 1908, Henrietta Swan Leavitt identified Cepheids as an important cosmic benchmark for the scaling of galactic and extragalactic distances after studying thousands of variable stars in the Magellanic Clouds. This discovery permitted a determination of the true luminosity of a Cepheid by observing its pulsation period. This relationship, in turn, enabled a determination of the distance to the variable star, by comparing its known luminosity to its observed brightness. The discovery of Leavitt gave astronomers their first "standard candle" with which to measure distances to faraway galaxies.

Harlow Shapley initially suggested that the cosmos was only made up of stars in the Milky Way system. However, Edwin Hubble detected Cepheid Variables in the Andromeda Galaxy in 1923 and 1924. By the early 1950s, it was known that the sun is a fairly typical star that is in no way special. This observation revealed a distance well beyond that of the Milky Way. Such findings led to the "cosmological principle", which indicates that the universe looks the same (on a large scale) no matter from where it is viewed. The cosmological principle is an important basis of big-bang cosmology.

Originally there was a debate between the Big Bang model and the Steady-State (static) model. The big bang model was subsequently supported by astronomical evidence with the discovery of a red-shift in 1929 by Hubble. The big bang model was further confirmed with the significant discovery by Arno Penzias and Robert Woodrow Wilson in 1964 of the cosmic microwave background radiation. The measurement of this background radiation proved crucial to this debate. The Big Bang model is particularly consistent with the background cosmic radiation having a black-body spectrum with a temperature of 2.725 K.

6.2   Observational Evidence

Advanced observational evidence in more modern times has significantly progressed. This evidence has included further measurements of the microwave background from the Cosmic Background Explorer (COBE), Wilkinson Microwave Anisotropy Probe (WMAP) and Planck satellites.

In addition, large galaxy red-shift surveys were conducted including the two-degree Field Galaxy Redshift Survey (2dFGRS) and the Sloan Digital Sky Survey (SDSS). Observations were also made of distant supernovae and gravitational lensing. These observations directly support predictions of the Lambda-CDM model. Further experimental support of Einstein's general theory of relativity included the announcement of the detection of gravitational waves in March 2014 by the Harvard–Smithsonian Center for Astrophysics. Gravitational waves themselves would be made of spin-2 *gravitons* like light waves are made of photons. Although gravitational waves have been detected, the graviton has not been observed since gravity is such a weak force.

Recent estimates of the make-up of the universe suggest that it includes: 4.9% atomic matter, 26.6% dark matter and 68.5% dark energy. Evidence of dark matter is derived from observed gravitational phenomena that include:

- Spiral galaxies are observed to spin too fast, where the gravitational pull from all stars within the galaxy is insufficient to counter the centripetal force that would stop the galaxies from flying apart. Moreover, the spectral lines of hydrogen far from the edge of the galaxy indicates a constant rotation speed at a far distance. Such observations can be explained by Newtonian dynamics only if there is a much greater mass than what is visible.
- A galaxy cluster, consisting of a hundred to a thousand galaxies, is bound together by gravity. The relationship between the speed of the galaxies and their separation allows the extraction of the mass of the galaxy. This estimated mass is a hundred times greater than the visible mass.
- Light bends as it passes by heavy masses that can result in a distorted image, which is known as *gravitational lensing*. Even small distortions allow for a determination of the mass of the cluster in the foreground, which exceeds that of visible matter.
- *Big bang nucleosynthesis* describes how light elements were formed in the early universe. The relative abundance of different elements depends on the total amount of baryon matter. This analysis indicates that the total amount of ordinary matter is just a few percent of the total energy density.
- The cosmic microwave background displays hot and cold regions of space. Such fluctuations provide the seeds for the later formation of clusters, galaxies and stars. This formation process cannot be achieved with only ordinary matter due to a coupling to photons, which causes

a pressure that suppresses gravitational collapse. On the other hand, dark matter does not interact with photons allowing for gravitational collapse to more effectively proceed. Moreover, ripples in the cosmic microwave background can arise from the inflation of a quantum field. When the universe expanded, quantum fluctuations were stretched from the microscopic scale to distances that span the visible universe so that hot and cold regions were frozen in place.

6.3 Evolution of the Universe

The Friedmann equation describes the expansion of the universe. This equation is derived considering the potential energy and kinetic energy of a test particle. It describes the evolution of the separation distance between two particles. Since the universe is homogeneous on a large scale, as a result of the cosmological principle, one can instead define a "comoving" system of coordinates. These coordinates are simply carried along with the expansion of the universe. Hence, with a uniform expansion, the real distance and comoving distance are simply proportional to one another, in which the proportionality constant between these two distances is termed the "scale factor of the universe".

Hubble discovered that galaxies are moving away (i.e., receding) from the Earth at speeds proportional to their distance. The velocities of the galaxies can be determined by their "red-shift". This was the first observational evidence that the universe is expanding thereby providing support for the Big Bang theory.

Problems of measuring distances to galaxies has complicated accurate measurements of the Hubble constant. The observed Hubble constant is normally parametrized as $H_0 = 100\ h$ km s^{-1} Mpc^{-1}, where the reported parameter h is a derived quantity from astrophysical measurements. Converting kilometres into megaparsecs, $H_0^{-1} = 9.77\ h^{-1} \times 10^9$ y. This inverse quantity is known as "Hubble time", where the uncertainty in the Hubble constant is reflected in h. This latter quantity provides an estimate for the age of the universe. Recent investigations from the Hubble telescope and Planck satellite indicate that $h = 0.7$ within a few percent.

It is currently believed that the expansion of the universe is accelerating based on studies of distant IA supernovae. This type of supernova occurs in binary systems in which one of the stars is a white dwarf. It is thought that the Universe has a geometry that is extremely close to a flat structure based on observations of the cosmic microwave background. Based on the value of the Hubble constant, one can estimate a "critical density" that arises for a flat universe. A dimensionless parameter called the "density parameter" Ω is simply the fraction of the actual density estimated from radiation, matter and energy contained in the universe to the critical density. The sum of all density parameters indicate that there is not enough matter in the universe to stop its expansion. The universe contains radiation with an energy density of $\Omega_{rad}h^2 = 2.47 \times 10^{-5}$. The combined relativistic photon and neutrino energy density is $\Omega_{rel}h^2 = 4 \times 10^{-5}$ (assuming neutrinos are massless). On the other hand, it is currently thought that the neutrinos have mass. Considering present limits for the mass of these particles, the neutrino density is estimated as $\Omega_\nu = 0.01$. In the process of nucleosynthesis that occurred in the early moments of the big bang, a soup of particles known as quarks and gluons condensed into protons and neutrons. The creation of these baryons make up 5% of the critical density. Anisotropy studies of the cosmic microwave background suggest a similar value for the baryon density.

Cosmic microwave background studies also show that dark matter has a value of $\Omega_{dm} = 0.27$. However, the precise composition of dark matter is not established. Dark matter is concentrated in galaxy halos. This conjecture is needed in order to explain the rotation curves of galaxies. On the other hand, the existence of dark energy arose from measurements of supernovas. It is an unknown form of energy that affects the largest scales of the universe. It has been postulated in order to account for the accelerating nature of the universe. The most recent measurements suggest that dark energy contributes 68% of the total energy in the observable universe today. As mentioned, the photons and neutrinos contribute a very small amount to the total energy content. Dark energy dominates the universe's mass-energy content because it is uniform across the space. The "cosmological constant" is one form of dark energy with a constant energy density filling space homogeneously. "Quintessence" is another form of dark energy which is a time-varying scalar field that is normally included in the cosmological constant.

Inflation is an important part of the Standard Cosmological Model since it can explain several problems of the hot big bang model. In particular, this approach can explain the extreme flatness of the universe, the horizon/communication problem of different regions in the universe, and the abundances of relic particles formed in its early creation.

The standard cosmological model suggests that the universe is 13.8 billion years old. It is expected to survive with an accelerated expansion

Table 6.1. Different stages of the universe's evolution.

Time (s)	Temperature (K)	Stage of Evolution
$t < 10^{-10}$	$T > 10^{15}$	Not known?
$10^{-10} < t < 10^{-4}$	$10^{12} < T < 10^{15}$	Strongly interacting free electrons, quarks, photons, neutrinos.
$10^{-4} < t < 1$	$10^{10} < T < 10^{12}$	Strongly interacting free electrons, quarks, photons, neutrinos.
$1 < t < 10^{12}$	$10^4 < T < 10^{10}$	Protons & neutrons form atomic nuclei. Free electrons, atomic nuclei, photons & neutrinos. Everything is strongly interacting except neutrinos.
$10^{12} < t < 10^{13}$	$3000 < T < 10^4$	As before. Universe is matter dominated.
$10^{13} < t_0$	$3 < T < 3000$	Atoms have formed from nuclei and electrons. Photons are no longer interacting & cool to form microwave background.

instead of re-collapsing. The different stages for the time evolution of the universe (taking $\Omega_0 = 0.3$ and $h = 0.7$) is given in Table 6.1, where t_0 indicates the present time. Evidence for the Standard Cosmological Model specifically includes:

(i) the observed expansion of the universe;
(ii) the predicted age of the universe;
(iii) the existence and thermal signature of the cosmic microwave background;
(iv) the formation of the light elements (e.g., deuterium, helium-3, lithium and helium-4) in the early creation of the universe through the process of nucleosynthesis;
(v) the ability of the model to predict observed galactic structures and the cosmic microwave background.

The recent launch of the James Webb Space Telescope provided recent images in 2022 of early stars in a globular cluster in a galaxy that were just 4 billion years old. This observation reveals to date some of the oldest stars known, considering the predicted age of the universe.

There have been many attempts over the years at constructing an ideal theory of gravity, although the simple tensor theory of general relativity has withstood many tests over time. Typical theories have typically involved scalar, tensor, scalar–tensor and vector–tensor fields. Scalar–tensor theories all contain at least one free parameter, whereas general relativity has none. In these theories, a vector field has both magnitude and direction while a scalar field only has a magnitude at every point in space–time. Tensors are generalized constructs of scalars and vectors. General vector–tensor theories are difficult to rule out as they can reduce to general relativity for certain parameter choices.

Motivations for recent alternative theories have been aimed at replacing cosmological constructs such as inflation, dark matter and dark energy. These theories also invoke a cosmological constant and have added scalar or vector potentials. John Moffat developed a MOdified Gravity (MOG), which is a non-symmetric theory of gravity based on a scalar–tensor–vector gravity (STVG) approach. It assumes a variable gravitational constant. Other theories assume that the speed of light may have been 30 orders of magnitude higher during the early moments of the big bang. In addition to a tensor field, MOG proposes the existence of a vector field, while replacing three constants of the theory with scalar field quantities. Far from a source of gravity, the STVG theory is stronger compared to the prediction of Newtonian gravity. However, at shorter distances, it is counteracted by a repulsive fifth force due to the presence of the vector field. The STVG theory is an alternative theory to the general theory of relativity. In particular, it has been successfully used to explain the rotation curves of galaxies, the mass profiles of galaxy clusters, and gravitational lensing effects without the need to invoke dark matter. Thus, importantly, it has been used to explain cosmological observations thereby eliminating the need for the presence of exotic dark matter that has not been directly observed nor identified. It further purports to explain anomalous effects in the cosmic microwave background data and accounts for the recently discovered acceleration of the universe. On a smaller scale, it reduces to the same predictions as general relativity.

However, more recent observations of extreme astrophysical systems, including a collision of two neutron stars in 2017 and a unique pulsar system in 2014, may have challenged predictions of such alternative gravitational theories as the modified Newtonian dynamics (MOND) and STVG approaches. The neutron star merger event yielded gravitational waves that were detected by the laser interferometer gravitational-wave observatory (LIGO) along with gamma ray bursts observed by the space-based Fermi satellite from the same location. These observations indicated that gravitational waves propagate at the speed of light to extremely high precision, which does not agree with predictions of a number of alternative gravitational theories. In addition, a tight compact triple system of stars, consisting of a pulsar and white dwarf orbiting one another along with a second white dwarf orbiting the pair, provided a further highly-accurate test of the strong equivalence principle of general relativity. This test used the high precision timing of the pulsar signals.

A "cosmological constant problem" is also an important issue in cosmology that describes the energy density of the universe. Here the vacuum energy density should gravitate in accordance with Einstein's theory of general relativity in the same manner as that for matter. This effect would cause the universe to slow down over time. In contrast, modern observations since 1998 suggest that the universe is expanding requiring a repulsive force to counteract gravity. To explain this result, the concept of "dark energy" was introduced that has a negative pressure causing the expansion of the universe to accelerate. The cosmological constant is identified as a uniform energy density that does not vary with time and location. However, a problem arises as the theoretical prediction of the magnitude of the cosmological constant is some 120 orders of magnitude larger than the observed value. Several solutions have been proposed that involve modifications to Einstein's general theory of relativity or modifications to the standard model but this issue persists as a challenging problem in modern cosmology.

Chapter 7

Black Holes

A black hole is a region of space–time where particles and/or electromagnetic radiation are unable to escape the strong gravitational forces generated by a sufficiently compact mass that can deform space–time. There is no escape from the boundary of a black hole, which is called the "event horizon". In accordance with the general theory of relativity, when an object crosses this boundary, the fate of an in-falling object is sealed as the space–time interval for the spatial position switches with the temporal one so that one cannot escape the black hole. Locally, however, there are no detectable features for an observer that crosses the event horizon.

At the centre of a black hole, a gravitational singularity with a zero volume as a single point occurs where the space–time curvature becomes infinite. With a rotating black hole, the singularity takes the form of a ring that lies in the plane of rotation of the black hole. The singular region regardless contains all the mass of the black hole. Observers falling into a non-rotating and uncharged black hole cannot avoid reaching the singularity at the centre of the black hole once they cross the event horizon. At the central singularity, an observer would be crushed with their mass added to the total mass of the black hole. Even before this dramatic event, they will have been torn apart by growing tidal forces.

The historical perspective in the development of black hole theory is described in Section 7.1. Characteristics of black hole theory are outlined in Section 7.2.

7.1 Historical Developments

The gravitational escape of objects was first considered in the 18th century by John Michell and Pierre-Simon Laplace. A solution of the field equations of general relativity for a point spherical mass was obtained by Karl Schwarzschild in 1916. A few months later, Johannes Droste provided the same solution for the point mass and described its properties. This solution described a so-called "Schwarzschild radius", where some terms in the Einstein field equations become infinite (i.e., singular). Arthur Eddington showed in 1924 that the singularity disappears with a change of coordinates, although it took until 1933 for Georges Lemaître to realize that this meant the singularity at the Schwarzschild radius was a "non-physical singularity" due to the coordinate system used.

In 1931, Subrahmanyan Chandrasekhar showed that a non-rotating body of electron-degenerate matter is unstable above a certain limiting mass (called the Chandrasekhar limit). This limit was calculated as 1.4 solar masses. Later, it was realized that a white dwarf, with a mass above the Chandrasekhar limit, will collapse into a stable neutron star. In 1939, Robert Oppenheimer and others predicted that neutron stars with a lower mass at the so-called Tolman–Oppenheimer–Volkoff (TOV) limit could collapse at only 0.7 solar masses by taking into consideration the Pauli exclusion principle. Accounting for neutron-neutron repulsion, as mediated by the strong force, this limit was raised to approximately 1.5 to 3.0 solar masses. Observations of the neutron star merger GW170817 was thought to have generated a black hole shortly afterwards. This observation refined the TOV limit to ∼2.17 solar masses. Oppenheimer and his co-authors interpreted the singularity at the boundary of the Schwarzschild radius as a boundary where time would stop from the point of view of an external observer. This situation occurs for the outside observer because any physical signal becomes infinitely red-shifted at the event horizon so that the signal is never able to reach the observer. Someone falling into a black hole, however, would not see time stop.

In the 1960s, significant theoretical work continued on black holes using the theory of general relativity. In this period, a number of general black hole solutions were found. In 1963, an exact solution for a rotating black hole was found by Roy Kerr. Ezra Newman found an axisymmetric solution for a rotating and electrically charged black hole in 1965. Werner Israel, Brandon Carter, Stephen Hawking and David Robinson further proposed the "no-hair theorem" for black holes. This theorem stated that a stationary black hole solution is completely described by three parameters in the Kerr–Newman metric based on the black hole's mass, angular momentum, and electric charge. These properties are apparent from outside of the black hole. A charged black hole will repel other like-charged objects.

The total mass inside a sphere containing a black hole can be found by using the gravitational analogue of Gauss's law for electromagnetism away from the black hole. The angular momentum (or spin) can also be measured far away with the occurrence of "frame dragging" where the rotation of a massive object distorts the space–time metric as a result of gravitoelectromagnetic currents, known as the "Lense-Thirring effect". Thus, a static (non-rotating) Schwarzschild black hole is characterized simply by its mass. A static but charged black hole is characterized both by its mass and charge, while a rotating black hole also depends on its angular momentum. A static black hole with an electric charge is called a "Reissner–Nordström black hole" while a rotating one is called a "Kerr black hole".

The belief of gravitational collapse for stars was sparked by the experimental discovery of pulsars by Jocelyn Bell Burnell in 1967. Pulsars were shown to be rapidly rotating neutron stars. Cygnus X-1 was the first black hole identified in 1971.

The mass of a black hole can take any positive value. The charge and angular momentum of a black hole are constrained by its mass. An event horizon will not occur for solutions of the general field equations that violate this requirement thereby giving rise to "naked singularities" that are able to be observed from outside of the black hole. For unphysical solutions of the theory, Roger Penrose suggested in 1969 that no naked singularities can exist in the universe. This hypothesis is known as the "cosmic censorship hypothesis". He also showed that once an event horizon forms, general relativity requires that a singularity will form within the black hole. This theorem is known as the "singularity theorem" for which he received the Noble Prize in 2020.

Early in the 1970s, Stephen Hawking also showed that the total area of the event horizon is analogous to the phenomenon of entropy in thermodynamics. In 1974, he also predicted that black holes are not entirely black but emit small amounts of thermal radiation, known as "Hawking radiation". By applying quantum field theory, the black hole will emit particles that display a perfect black body spectrum. Thus, black holes are expected to shrink and evaporate over time with a loss of mass with an emission of photons and other particles. The temperature of this thermal spectrum is proportional to the surface gravity of the black hole and inversely proportional to its mass. As such, large black holes emit less radiation than smaller ones. This temperature is on the order of a billionth of a degree Kelvin for stellar black holes so that it is too low to be detected.

It is far less than the temperature of the cosmic microwave background at 2.7 K. Larger black holes receive more mass from the cosmic microwave background than they emit through Hawking radiation and thus can grow in size. Hawking further revealed that a number of cosmological solutions that describe the Big Bang theory possess singularities.

The Kerr solution, the no-hair theorem, and the laws of black hole thermodynamics provide simple physical properties of black holes. Nevertheless, the appearance of singularities in general relativity indicates a possible breakdown of this classical theory. At extremely high densities at the location of the singularity, a combined theory that encompasses both quantum effects as well as gravity needs to account for particle interactions at this small scale. Such a theory has not yet been realized.

The LIGO Scientific Collaboration and Virgo collaboration announced the detection of gravitational waves on February 11, 2016, which confirmed predictions of Einstein's general theory of relativity. The observed gravitational waves were produced by a merger of several black holes. The image of a black hole was constructed from observations made with the Event Horizon Telescope (EHT) in 2017. These images were published in April 2019 and revealed a super-massive black hole in the galactic centre of Messier 87. The nearest black hole is found at a distance around 1,500 light-years away. It is thought that there are hundreds of millions of black holes but these do not emit radiation and can only be detected by gravitational lensing.

7.2   Characteristics of Black Holes

Black holes result when massive stars collapse at the end of their life cycle. After formation, they can grow by absorbing mass from their surroundings. Super-massive black holes (with a mass of millions of solar masses) are believed to exist at the centre of galaxies. These objects can form by absorbing matter from other stars or by merging with other black holes. The presence of a black hole can be identified by its interaction with stellar objects with the observance of in-falling matter into the black hole. Due to conservation of angular momentum, gas falling into the strong gravitational well of a black hole will typically form a disk-like structure around the object. Hence, a rotating disk of matter is formed as an external "accretion disk" around massive black holes. The gas in the inner accretion disk orbits at very high speeds given its proximity to the compact object. The resulting friction heats the inner disk to temperatures at which vast amounts of electromagnetic radiation (mainly X-rays) are emitted.

Quasars are bright objects believed to be the accretion disks of super-massive black holes and are therefore powered by black holes. The quasars have extremely luminous galactic cores as gas and dust fall into a super-massive black hole thereby producing electromagnetic radiation. Stars passing too close to a super-massive black hole can even be swallowed up by it. The orbit of stars can be used to infer the mass and location of a black hole. Numerous candidate black holes as part of a binary system have been identified. In fact, a radio source known as Sagittarius A*, at the centre of the Milky Way galaxy, is thought to contain a super-massive black hole.

A rotating black hole is surrounded by a region of space–time called the "ergosphere". As a consequence of frame-dragging, the rotating mass will "drag" along the space–time immediately surrounding the black hole. Thus, any object near the rotating mass will tend to move in the direction of rotation. In fact, this effect is so strong that near the event horizon of a black hole, an object would have to move faster than the speed of light in the opposite direction in order to remain stationary.

Astrophysical black holes are not known or expected to carry charge. They retain the nearly neutral charge of the star given the relatively large strength of the electromagnetic force. On the other hand, a black hole can rotate given the angular momentum of the original collapsing star. However, if the rotation is slow, a Schwarzschild solution can be used as a reasonable approximation.

As mentioned, an event horizon is a defining feature of a black hole through which matter and light can pass only inward towards the mass of the black hole. Nothing can escape from inside through the event horizon. General relativity predicts that the large mass deforms space–time so that paths taken by particles bend toward the mass itself. This deformation becomes so strong at the event horizon that there are no paths that lead away from the black hole. Clocks near a black hole also appear to tick more slowly than those farther away as seen by a distant observer. Thus, with so-called "gravitational time dilation", an object falling into a black hole appears to slow as it approaches the event horizon, taking an infinite time to reach it. All processes slow down from the viewpoint of a fixed observer outside the black hole. Light emitted by the object appears redder and dimmer due to the "gravitational red-shift". Eventually, the falling object fades from view. On the other hand, observers that fall into a black hole do not notice any of these effects as they cross the event horizon in which their own clocks appear to tick normally.

There is a direct correspondence of "black body mechanics" with ther-modynamics. In this direct analogy, the mass of a black hole corresponds to energy, the surface gravity corresponds to temperature and the area of the event horizon corresponds to entropy. Interestingly, the entropy of a black hole scales with its area rather than with its volume. Since entropy normally scales with the volume of a system, this property led Gerard t'Hooft and Leonard Susskind to suggest a so-called "holographic principle". In this principle, anything that happens in a volume of space–time can be described by data on the boundary of that volume.

A black hole has only a few internal parameters (i.e., total mass, charge, and angular momentum). Information about matter that falls into a black hole is lost. This information exists inside the black hole, which is inaccessible from the outside, but represented on the event horizon as per the holographic principle. However, since black holes can slowly evaporate by emitting Hawking radiation, information about the in-falling matter that formed the black hole is lost forever. In quantum mechanics, such a loss of information violates the "unitarity" property of quantum mechanics needed to describe the progress and time evolution of a system. In particular, it is not possible for a pure quantum state to evolve into a mixed state. The Hawking radiation emitted is thermal and therefore random. Hence, a pure state falls into a black hole and is emitted as a mixed state involving a non-unitary evolution. Another way of stating this problem is that according to quantum field theory in curved space–time, a single emission of Hawking radiation involves two mutually entangled particles. A black hole that is formed in a finite time in the past will evaporate away in a finite time. The black hole will therefore only emit a finite amount of information. There will eventually be a time where an outgoing particle must be entangled with all the Hawking radiation the black hole had previously emitted. This occurrence creates a paradox. Like any quantum mechanical system, the outgoing particle cannot be fully entangled with two other systems at the same time. The outgoing particle appears to be entangled both with the in-falling particle and, independently, with past Hawking radiation. Hence, several time-tested principles may be challenged including Einstein's equivalence principle, and unitarity or local quantum field theory. As general relativity involves a semi-classical calculation of black-hole mechanics, this situation suggests that a full quantum gravitational treatment is needed to solve the loss of information paradox.

Chapter 8

Epilogue

There still remain important questions in theoretical physics yet to be resolved or new theories to be advanced and developed. Such phenomena still not understood include the following:

(i) A rigorous quantum field theory, with interactions as constructed in four-dimensional space–time, does not exist to replace the current theory that must resort to perturbative methods of solution.

(ii) A complete theory of fundamental interactions does not exist. For instance, general relativity is a classical theory. A quantum theory of gravity would require a fully-consistent and unified theory of quantum mechanics and general relativity that is supported by experimental and observational evidence. In addition, a full unification of the electroweak force with the strong force (i.e., quantum chromodynamics theory) is needed for all particle energies. Currently, an ad hoc theory has been developed where the electroweak and strong forces act separately in the standard model and are therefore not truly unified. Gravity remains a separate stand-alone classical theory as one of the four fundamental forces of nature. As a hierarchical problem, it is not clear why gravity is such a weak force compared to the other fundamental forces of nature.

(iii) The standard model does not account for neutrino oscillations and their non-zero masses. In particle physics, an explanation for the discrepancy between the measured and theoretical value of the anomalous magnetic dipole moment is needed. It is also not understood why there are three generations of quarks and leptons in particle physics.

(iv) The standard model does not contain any viable particle candidates for dark matter that possess the required properties as deduced from observational cosmology. In addition, there is no specific theory accounting for the accelerating nature of the expansion of the universe, as purposed with the concept of "dark energy" in cosmology. It is not clear why the mass of the quantum vacuum does not directly affect the expansion of the universe.

(v) There is an observed matter-antimatter asymmetry in nature. This asymmetry includes an imbalance in the baryonic matter as experienced in everyday life as well as antibaryonic matter existing in the observable universe. Neither the standard model of particle physics nor the general theory of relativity can explain this asymmetry. The Big Bang, in reality, should have produced equal amounts of matter and antimatter. Thus, some physical laws must have acted differently during the evolution of the universe to explain this imbalance of matter and antimatter in the baryogenesis process.

(vi) In the current formulation of chromodynamics theory, a violation of charge-parity(CP)-symmetry in strong interactions can occur. Nevertheless, no violation has ever been observed in any experiments involving this type of interaction. Thus, the observed conservation of CP in particle physics suggest a "CP-problem" in the present standard model. Also, it is not understood why magnetic monopoles do not exist in nature, which are important for understanding the dynamics of strongly coupled gauge fields such as in quantum chromodynamics.

(vii) The physical nature of the hypothetical inflaton scalar field, which gave rise to cosmic inflation in the early universe, is not known.

(viii) There is no physical evidence to show that black holes irradiate thermal energy, which has only been suggested on theoretical grounds. With black hole evaporation, there is a loss of information that occurs giving rise to an information paradox for evaporating black holes. Black hole evaporation is specifically predicted in accordance with classical theories suggesting again the need for a quantum gravity theory. A quantum gravity theory is needed to explain phenomena near the singularity of a black hole with extreme densities where gravity is so strong that space–time itself breaks down catastrophically.

Appendices:
Supplemental Mathematics

A   Classical Theory

Lagrangian and Hamiltonian formalisms in Appendix A.1 provide a means with alternative mathematical ways to describe the dynamics of a system that arise in classical mechanics. Lagrangian mechanics was named after Joseph–Louis Lagrange (1736–1813), while Hamiltonian mechanics was named after William Rowan Hamilton (1805–1865). These latter approaches are reformulations of Newtonian mechanics. The functional relationship between the Lagrangian $L(q, \dot{q}, t)$ (i.e., with generalized coordinates and velocities with time as a parameter) and Hamiltonian $H(q, p, t)$ (with generalized coordinates, momenta and time) is given by a Legendre transformation. This type of transformation is used frequently in thermodynamics. It allows a change in basis from the $(q, \dot{q}, t)$ set of variables to the (q, p, t) set. The Hamiltonian equations constitute a set of first-order equations of motion replacing the Lagrange equations which are of second order. These formulations have many applications in theoretical physics. Symmetries of the Lagrangian leads to conservation laws in accordance with Noether's theorem (see Appendix A.1.1).

The Lagrangian plays an important role in electromagnetism, general relativity and classical field theory. For example, as shown in Appendix A.2, the Lagrangian density can be used to derive Maxwell's equations. Furthermore, as shown in Appendix B, the field equations for general relativity can also be developed from a generalized Lagrangian density function. The Lagrangian density in Eq. (D.17) is important for quantum electrodynamics in the Standard Model of Appendix D.

Hamiltonian mechanics are used as a basis for "canonical quantization" in Appendix C. Here formal rules are applied to replace classical observables with quantum mechanical operators. For example, as demonstrated in Appendix C.1, the Hamiltonian operator can be expressed for a single particle constrained in a potential for a given position and time. This operator is used as the basis for the Schrödinger equation.

The Maxwell equations in Appendix A.2 provide an overall formalism that encompasses the classical theory of electrodynamics. These equations importantly demonstrate that all observers measure the same speed of light for electromagnetic wave propagation, which underpins the special theory of relativity. The Maxwell equations describe how electric charges and electric currents create electric and magnetic fields. They also show how an electric field can give rise to a magnetic field.

A.1 Classical Mechanics

The Lagrangian formulation is derived from the principle of least action (Appendix A.1.1). In contrast, the Hamiltonian formulation is based on Hamilton's principle of stationary total energy (Appendix A.1.2). The Lagrangian is a function that describes the kinetic energy of a system minus its potential energy. The Hamiltonian is a function that describes the total energy of the system. Both formulations are related to one another through a Legendre transformation. Both tools are useful for understanding the dynamics of a physical system. They are used depending on which one is more specific to the given problem.

A.1.1 *Lagrangian*

The Lagrangian L is used to derive the equations of motion for a system. Here the action S is defined as the integral of the Lagrangian over time for the development of a system from an initial time to a final time:

$$S = \int_{t_1}^{t_2} L(q(t), \dot{q}, t) dt \tag{A.1}$$

The parameters q and $\dot{q}$ are generalized coordinates and the dot pertains to a derivative with respect to the independent variable time t. The endpoints of the evolution are fixed as $q_1 = q(t_1)$ and $q_2 = q(t_2)$. The *true* evolution of a system (i.e., equations of motion) is one in which the action is stationary. The "Euler–Lagrange equation" can be derived by applying the principle of least action from Eq. (A.1):

$$\delta S = \delta \int_{t_1}^{t_2} L(q, \dot{q}, t)dt = \int_{t_1}^{t_2} \left(\frac{\partial L}{\partial q}\delta q + \frac{\partial L}{\partial \dot{q}}\delta \dot{q} \right) dt \qquad (A.2a)$$

$$= \int_{t_1}^{t_2} \left(\frac{\partial L}{\partial q}\delta q + \frac{\partial L}{\partial \dot{q}}\frac{d(\delta q)}{dt} \right) dt = 0 \qquad (A.2b)$$

Integrating the last term of Eq. (A.2b) by parts on letting:

$$u = \frac{\partial L}{\partial \dot{q}} \quad \text{and} \quad dv = d(\delta q)$$

$$\qquad\qquad\qquad\qquad\qquad\qquad\qquad (A.3)$$

$$du = \frac{d}{dt}\left(\frac{\partial L}{\partial \dot{q}} \right) dt \quad \text{and} \quad v = \delta q$$

yields

$$\int_{t_1}^{t_2} \left(\frac{\partial L}{\partial \dot{q}}\frac{d(\delta q)}{dt} \right) dt = \int_{t_1}^{t_2} \frac{\partial L}{\partial \dot{q}}d(\delta q) = \left[\frac{\partial L}{\partial \dot{q}}\delta q \Big|_{t_1}^{t_2} - \int_{t_1}^{t_2}\left(\frac{d}{dt}\frac{\partial L}{\partial \dot{q}}dt \right)\delta q \right]$$

$$\qquad\qquad\qquad\qquad\qquad\qquad\qquad (A.4)$$

Inserting Eq. (A.4) into Eq. (A.2b) gives:

$$\delta S = \frac{\partial L}{\partial \dot{q}}\delta q \Big|_{t_1}^{t_2} + \int_{t_1}^{t_2}\left(\frac{\partial L}{\partial q} - \frac{d}{dt}\frac{\partial L}{\partial \dot{q}} \right)\delta q \, dt = 0 \qquad (A.5)$$

Since only the path is varying and not the endpoints, $\delta q(t_1) = \delta q(t_2) = 0$ and Eq. (A.5) becomes

$$\delta S = \int_{t_1}^{t_2}\left(\frac{\partial L}{\partial q} - \frac{d}{dt}\frac{\partial L}{\partial \dot{q}} \right)\delta q \, dt = 0 \qquad (A.6)$$

Hence, for an arbitrary small change δq for a stationary action the integrand itself must equal zero, resulting in the Euler–Lagrange equation:

$$\boxed{\frac{\partial L}{\partial q} - \frac{d}{dt}\frac{\partial L}{\partial \dot{q}} = 0} \qquad (A.7)$$

Example A.1. The Euler–Lagrange equation can be used to derive Newton's second law of motion. Let the generalized coordinate be position for the dependent variable so that $q(t) \rightarrow x(t)$ and $q(t) \rightarrow \dot{x}(t)$ where $\dot{x} = dx/dt$ is the velocity. The Euler–Lagrange equation in Eq. (A.7) can therefore be written as:

$$\frac{d}{dt}\frac{\partial L}{\partial \dot{x}} - \frac{\partial L}{\partial x} = 0 \qquad \text{(A.8)}$$

In this case, the Lagrangian can be defined as the difference between the kinetic energy T and potential energy V of a system:

$$L = T - V \qquad \text{(A.9)}$$

The kinetic energy of a particle of mass m is:

$$T = \frac{1}{2}m\dot{x}^2 \qquad \text{(A.10)}$$

The negative gradient of the potential energy $V(x)$ is defined as the force F:

$$F(x) = -\frac{dV(x)}{dx} \qquad \text{(A.11)}$$

Thus, substituting Eq. (A.9) into Eq. (A.8) and using Eq. (A.10) and Eq. (A.11) for the kinetic and potential energies, respectively, yields:

$$\frac{d}{dt}(m\dot{x}) - \frac{\partial(-V)}{\partial x} = m\ddot{x} - F(x) = 0 \Rightarrow F = ma \qquad \text{(A.12)}$$

This result is Newton's second of law of motion where a is the acceleration such that $a = \ddot{x}$.

Example A.2. Noether's theorem is an important theorem utilized in classical mechanics, quantum mechanics and quantum field theory. As mentioned in Section 1.1, this theorem establishes a link between continuous symmetries and conservation laws such that if a system has continuous symmetry, then there is a corresponding quantity whose values are conserved over a period of time.

Consider an illustrative example of Noether's theorem for the simple case of a coordinate transformation with a particle moving on a line with Lagrangian $L(q, \dot{q})$ where q is the particle position and $\dot{q} = dq/dt$ is its velocity. From example A.1, the momentum and force on the particle are defined as:

$$p = \frac{\partial L}{\partial \dot{q}} \qquad \text{(A.13)}$$

and

$$F = \frac{\partial L}{\partial q} \qquad \text{(A.14)}$$

Moreover, as follows from the Euler-Lagrange equations in example A.1:

$$\dot{p} = F \tag{A.15}$$

If the Lagrangian has a symmetry, it will not change when applying some one-parameter of transformations such as, for example, a translation in q to a new position $q(s)$. In this case:

$$\frac{d}{ds} L\left(q(s), \dot{q}(s)\right) = 0 \tag{A.16}$$

Now, letting $C = p\dfrac{dq(s)}{ds}$ then the time derivative using the product rule is:

$$\dot{C} = \dot{p}\frac{dq(s)}{ds} + p\frac{d\dot{q}(s)}{ds} \tag{A.17}$$

Using the equation of motion for the particle in Eq. (A.14), along with the definition of the momentum and its time derivative in Eq. (A.13) and Eq. (A.15) to rewrite p and $\dot{p}$ in Eq. (A.17), one obtains:

$$\dot{C} = \frac{\partial L}{\partial q}\frac{dq(s)}{ds} + \frac{\partial L}{\partial \dot{q}}\frac{d\dot{q}(s)}{ds} \tag{A.18}$$

By the chain rule, Eq. (A.18) can be written as

$$\dot{C} = \frac{d}{ds} L\left(q(s), \dot{q}(s)\right) = 0 \tag{A.19}$$

It follows that the expression in Eq. (A.19) equals zero from Eq. (A.16). Hence, the quantity C itself is a constant, which is therefore a conserved quantity.

In fact, in physics, there are many symmetries with conservation laws that include, for example:

- Translation in space for conservation of linear momentum.
- Translation in time for conservation of energy.
- Rotational symmetry for conservation of angular momentum.
- Gauge transformation for conservation of the electric, weak and colour charge.

A.1.2   *Hamiltonian*

The Hamiltonian describes a system in terms of generalized coordinates $\{q\}$ and momenta $\{p\}$. The Hamiltonian function H is defined in terms of the

Lagrangian L through the "Legendre transformation" of L for n generalized coordinates:

$$H(p, q, t) = \sum_{i=1}^{n} p_i \dot{q}_i - L(q, \dot{q}, t) \tag{A.20}$$

where the Lagrangian is a general function of the coordinates and velocities. Thus, the change in the Hamiltonian δH is given by

$$\delta H = \sum_{i=1}^{n} (p_i \delta \dot{q}_i + \dot{q}_i \delta p_i) - \delta L$$

$$= \sum_{i=1}^{n} \left(p_i \delta \dot{q}_i + \dot{q}_i \delta p_i - \frac{\partial L}{\partial q_i} \delta q_i - \frac{\partial L}{\partial \dot{q}_i} \delta \dot{q}_i \right) - \frac{\partial L}{\partial t} dt \tag{A.21}$$

Defining the conjugate momentum as

$$p_i = \frac{\partial L}{\partial \dot{q}_i} \tag{A.22}$$

the Euler–Lagrange equation in Eq. (A.7) can be rewritten as

$$\dot{p}_i = \frac{\partial L}{\partial q_i} \tag{A.23}$$

Using Eq. (A.22), the first and last terms cancel in the parenthesis exactly in Eq. (A.21), yielding:

$$\delta H = \sum_{i=1}^{n} \left(\dot{q}_i \delta p_i - \frac{\partial L}{\partial q_i} \delta q_i \right) - \frac{\partial L}{\partial t} dt \tag{A.24}$$

This latter equation can be directly compared to the general rule for a small change in a function of several variables:

$$\delta H = \sum_{i=1}^{n} \left(\frac{\partial H}{\partial p_i} \delta p_i + \frac{\partial H}{\partial q_i} \delta q_i \right) + \frac{\partial H}{\partial t} dt \tag{A.25}$$

yielding $\frac{\partial H}{\partial p_i} = \dot{q}_i$, $\frac{\partial H}{\partial q_i} = -\frac{\partial L}{\partial q_i}$ and $\frac{\partial H}{\partial t} = -\frac{\partial L}{\partial t}$. Finally, using the rewritten Euler–Lagrange equation in Eq. (A.23) gives Hamilton's equations:

$$\boxed{\frac{\partial H}{\partial p_i} = \dot{q}_i, \frac{\partial H}{\partial q_i} = -\dot{p}_i \quad \text{and} \quad \frac{\partial H}{\partial t} = -\frac{\partial L}{\partial t}} \tag{A.26}$$

Example A.3. The Hamiltonian equations can be alternatively used to derive Newton's second law of motion. Computing the canonical momenta from Eq. (A.22):

$$p = \frac{\partial L}{\partial \dot{x}} = m\dot{x} \tag{A.27}$$

This equation can be used in the definition of the Hamiltonian in Eq. (A.20):

$$H = p\dot{x} - L = m\dot{x}^2 - \frac{1}{2}m\dot{x}^2 + V = \frac{1}{2}m\dot{x}^2 + V \tag{A.28}$$

Equation (A.28) corresponds to the total energy of the system. The Hamiltonian equation of motion follows from the second equation in Eq. (A.26), where using Eq. (A.28):

$$\dot{p} = \frac{dp}{dt} = -\frac{\partial H}{\partial x} = -\frac{\partial V(x)}{\partial x} \tag{A.29}$$

The negative gradient of the potential energy $V(x)$ is defined as the force F in Eq. (A.11). Hence, the same result that $F = ma$ is obtained as in example A.1 using the Lagrangian function since $\dot{p} = ma$.

A.2   Maxwell's Equations

Maxwell's equations describe the behaviour of electric and magnetic fields in four separate equations. These equations encompass previous laws of classical electromagnetism that are mathematically described as:

(i) *Gauss's law for electric fields* where the electric flux through any closed surface is proportional to the charge enclosed within the surface:

$$\nabla \cdot \mathbf{E} = \frac{\rho}{\epsilon_0} \tag{A.30}$$

where $\mathbf{E}$ is the electric field and ϵ_0 is the permittivity of free space.

(ii) *Gauss's law for magnetic fields* indicates that no magnetic monopoles exist, such that the divergence of the magnetic field is always zero:

$$\nabla \cdot \mathbf{B} = 0 \tag{A.31}$$

where $\mathbf{B}$ is the magnetic field.

(iii) *Faraday's law of induction* where a changing magnetic field induces an electric field that generates an electromotive force:

$$\nabla \times \mathbf{E} = -\frac{\partial \mathbf{B}}{\partial t} \tag{A.32}$$

(iv) *Ampère's law (including a correction by Maxwell for the displacement current)* that describes the relationship between electric currents and the magnetic field they create:

$$\nabla \times \mathbf{B} = \mu_0 \mathbf{J} + \mu_0 \epsilon_0 \frac{\partial \mathbf{E}}{\partial t} \tag{A.33}$$

where $\mathbf{J}$ is the current density and μ_0 is the permeability of free space.

Consider Eq. (A.30) to Eq. (A.33) in a region with no charges ($\rho = 0$) and no currents ($\mathbf{J} = 0$) such as in a vacuum. Maxwell's equations then reduce to the homogeneous and inhomogeneous pairs:

$$\nabla \cdot \mathbf{E} = 0, \ \ \nabla \times \mathbf{E} = -\frac{\partial \mathbf{B}}{\partial t}$$

$$\nabla \cdot \mathbf{B} = 0, \ \ \nabla \times \mathbf{B} = \mu_0 \epsilon_0 \frac{\partial \mathbf{E}}{\partial t} \tag{A.34}$$

Taking the curl of the curl equations in Eq. (A.34) for the inhomogeneous equations, using the identity $\nabla \times (\nabla \times \mathbf{A}) = \nabla(\nabla \cdot \mathbf{A}) - \nabla^2 \mathbf{A}$, and substituting in the homogeneous equations into this identity along with the original inhomogeneous equations, yields the two standard wave equations:

$$\frac{1}{c^2}\frac{\partial^2 \mathbf{E}}{\partial t^2} - \nabla^2 \mathbf{E} = 0 \ \ \text{and} \ \ \frac{1}{c^2}\frac{\partial^2 \mathbf{B}}{\partial t^2} - \nabla^2 \mathbf{B} = 0 \tag{A.35}$$

where $c = 1/\sqrt{\mu_0 \epsilon_0}$, which is evaluated as the speed of light in free space. Thus, electromagnetic waves importantly travel at a constant speed of light in vacuum, where all observers measure the same speed as postulated in special relativity.

The Maxwell equations can be cast into a "covariant form". In this form, the Maxwell equations and Lorentz force are invariant under a Lorentz transformation (see Appendix B) using the formalism of special relativity for a flat space–time coordinate system (i.e., Minkowski metric tensor $\eta_{\mu\nu} = \eta^{\mu\nu}$ in Appendix B.2.2). For the covariant formalism, the "electromagnetic field tensor", which is antisymmetric, is defined as

$$F^{\mu\nu} \equiv \eta^{\mu\alpha} F_{\alpha\beta} \eta^{\beta\nu} = \partial^\mu A^\nu - \partial^\nu A^\mu$$

$$= \begin{pmatrix} 0 & -E_x/c & -E_y/c & -E_z/c \\ E_x/c & 0 & -B_z & B_y \\ E_y/c & B_z & 0 & -B_x \\ E_z/c & -B_y & B_x & 0 \end{pmatrix} \tag{A.36}$$

Here $A^\nu = (\phi/c, \mathbf{A})$ is a four-vector whose time component is the scalar potential ϕ and the spatial component is the usual vector potential $\mathbf{A}$.

The four-gradient is defined as $\partial^\nu \equiv \frac{\partial}{\partial x_\nu} = \left(\frac{1}{c}\frac{\partial}{\partial t}, \nabla\right)$, where the four-displacement is defined as $x^\alpha \equiv (ct, \mathbf{x}) = (ct, x, y, z)$.

Without sources, the homogeneous Maxwell's equations for Faraday's law of induction and Gauss' law for magnetism combine using the electromagnetic field tensor:

$$\boxed{\partial^\sigma F^{\mu\nu} + \partial^\mu F^{\nu\sigma} + \partial^\nu F^{\sigma\mu} = 0} \tag{A.37}$$

The two inhomogeneous Maxwell's equations for Gauss' law for electric fields and Ampère's law further combine into:

$$\boxed{\partial_\alpha F^{\alpha\beta} = \mu_0 J^\beta} \tag{A.38}$$

where the contravariant four-vector combines the electric charge ρ and electric current density $\mathbf{J}$ such that $J^\alpha = (c\rho, \mathbf{J})$.

The Lagrangian density for classical electrodynamics is composed of two components, a field component and a source component:

$$\mathcal{L} = \mathcal{L}_{\text{field}} + \mathcal{L}_{\text{int}} = -\frac{1}{4\mu_0} F^{\alpha\beta} F_{\alpha\beta} - A_\alpha J^\alpha \tag{A.39}$$

Example A.4.   The Lagrangian density in Eq. (A.39) leads to the inhomogeneous Maxwell's equation in Eq. (A.38) if the potential A_α is varied but the density is constant. Thus, for the given action (see Appendix A.1.1):

$$S = \int d^4x \left(-\frac{1}{4\mu_0} F^{\alpha\beta} F_{\alpha\beta} - A_\alpha J^\alpha\right) \tag{A.40}$$

a least action results with a vanishing variation $\delta S = 0$:

$$\delta S = \int d^4x \left(-\frac{1}{4\mu_0}(\delta F^{\alpha\beta})F_{\alpha\beta} - \frac{1}{4\mu_0}F^{\alpha\beta}(\delta F_{\alpha\beta}) - \delta A_\alpha J^\alpha\right)$$

$$= \int d^4x (\partial_\alpha F^{\alpha\beta} - \mu_0 J^\beta)\delta A_\beta = 0 \tag{A.41}$$

The last line in Eq. (A.41) is obtained using the antisymmetric property of the electromagnetic tensor and the relation $F_{\alpha\beta} = \partial_\alpha A_\beta - \partial_\beta A_\alpha$ as follows from Eq. (A.36). Additional operations include the relabelling of repeated (dummy) indices, raising and lowering of indices and an integration by parts [McMahon, 2008]. Since the variation is arbitrary, δA_β cannot vanish so that the integrand must therefore be zero everywhere leading to the Maxwell's equations in Eq. (A.38).

B Theory of Relativity

Supplemental mathematical derivations are given for important equations in special relativity (Appendix B.1) and general relativity (Appendix B.2). This treatise includes formulae for the Lorentz transformation in special relativity, which leads to such effects as "time dilation" and "length contraction". In addition, an assessment of the "total relativistic energy" is given. The derivation of the general field equations in general relativity are also presented. The linearization of the field equations yields the Newtonian limit for gravitation theory as well as the prediction of gravitational waves. Examples for the validation of relativity theory are also presented in Appendix B.3.

B.1 Special Relativity

B.1.1 *Lorentz transformation*

The Lorentz transformation for a Cartesian coordinate frame in space-time can be derived assuming a linear transformation from one inertial frame of reference (x, y, z, t) to another (x', y', z', t') for an event coinciding at $t = t' = 0$. Here the primed frame moves at a constant velocity v relative to the unprimed frame along the x-direction. In this derivation, it is implicitly assumed that all observers measure the same speed of light c. As such, it can be written [McMahon, 2006]:

$$t' = \gamma \left(t - vx/c^2 \right), \quad x' = \gamma \left(x - vt \right), \quad y' = y \quad \text{and} \quad z' = z \tag{B.1}$$

where $\gamma = \frac{1}{\sqrt{1-\beta^2}}$ and $\beta = v/c$. In the case that $v \ll c$, Eq. (B.1) reduces to the simple "Galilean transformation":

$$t' = t, \quad x' = (x - vt), \quad y' = y \quad \text{and} \quad z' = z \tag{B.2}$$

The Lorentz formalism affects time dilation and length contraction of a moving object, as well as the relativistic energy of a moving particle.

B.1.2 Time dilation

Consider the same frame configuration as for the derivation of the Lorentz transformation. An observer in the primed system measures an interval of time $\Delta t'$ at the same point in the rest frame. As seen by the observer in the other frame, $\Delta t = \gamma \Delta t'$. Each observer will therefore measure the time interval between ticks of a moving clock to be longer by a factor γ compared to the time interval between ticks of their own clock.

B.1.3 Length contraction

Again consider the same frame configuration in which a rod is at rest in the unprimed system that is aligned along the x-axis with a length Δx. In the primed system, the rod is seen to move with a velocity $-v$ so that its length must be measured by taking two simultaneous measurements at ($\Delta t' = 0$). An inverse Lorentz transform indicates $\Delta x = \gamma \Delta x'$. As such, observers measure the distance between the end points of a moving rod to be shorter by a factor $1/\gamma$ than the end points of the identical rod at rest in their own frame.

B.1.4 Relativistic energy

The total relativistic energy E is related to the rest mass energy m_0:

$$E \equiv mc^2 \equiv m_0 \gamma c^2 \tag{B.3}$$

The invariant mass (or rest mass) of the particle m_0 is measured in the instantaneous rest frame of the particle. Also v is the velocity of the particle moving with respect to an observer at rest. A Lorentz invariant is $E^2 - p^2 c^2 = m_0 c^4$, where the relativistic momentum is $p = m_0 c \beta \gamma$. Thus, if $pc \ll m_0 c^2$, then

$$E = m_0 c^2 \sqrt{1 + \frac{p^2 c^2}{m_0^2 c^4}} = m_o c^2 \left(1 + \frac{1}{2} \frac{p^2 c^2}{m_0^2 c^4} \cdots \right) = m_0 c^2 + \frac{1}{2} \frac{p^2}{m_0}$$

The last term is the non-relativistic kinetic energy.

B.2   General Relativity

The field equations of general relativity can be derived using a technique to find the shortest route or distance on a surface. In this process, one searches for trajectories that minimize a so-called "action", i.e., the shortest path length for the classic equations of motion. This method for general relativity was proposed independently by David Hilbert in 1915. On a flat surface, the shortest distance between two points is a straight line, while on a curved surface it is a "geodesic". In particular, on a sphere, a geodesic is simply a great circle route.

For instance, using the Pythagorean theorem (see Fig. B.1), the small distance element on a flat surface is [Lewis *et al.*, 2022]:

$$ds = [(dx)^2 + (dy)^2]^{1/2} \qquad (B.4)$$

Hence, the total distance assuming a starting point of (x_1, y_1) and an ending point of (x_2, y_2) may be written as

$$S = \int_{x_1,y_1}^{x_2,y_2} ds = \int_{x_1,y_1}^{x_2,y_2} [(dx)^2 + (dy)^2]^{1/2} \qquad (B.5)$$

This result is further generalized where the shortest distance in four dimensional space–time is a geodesic, which is a basic construct of general relativity. Thus, one can use a metric $g_{\mu\nu}$ for four-dimensional space–time

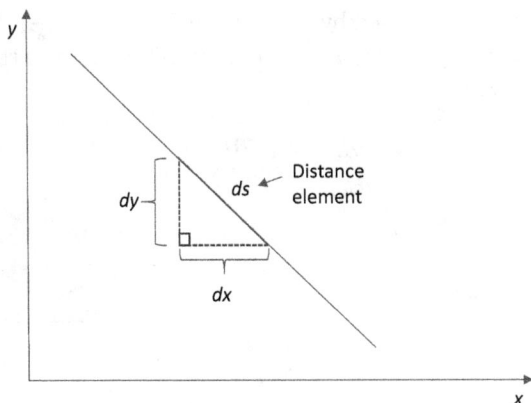

Figure B.1.   Schematic of a distance element in flat space.

to account for any type of surface and for any general coordinate system:

$$ds^2 = g_{\mu\nu}dx_\mu dx_\nu \tag{B.6}$$

For instance, to represent Eq. (B.4) with two dimensions (that is, with indices $\mu = 1, 2$ and $\nu = 1, 2$), Eq. (B.6) can be written in the following component form using the Einstein convention of summing on repeated indices [Lieber and Lieber, 1966]:

$$ds^2 = g_{11}dx_1 \cdot dx_1 + g_{12}dx_1 \cdot dx_2$$
$$+ g_{21}dx_2 \cdot dx_1 + g_{22}dx_2 \cdot dx_2 \tag{B.7}$$

where g for a flat (Euclidean) surface is the 2×2 matrix: $\begin{bmatrix} g_{11} & g_{12} \\ g_{21} & g_{22} \end{bmatrix} = \begin{bmatrix} 1 & 0 \\ 0 & 1 \end{bmatrix}$. Moreover, Eq. (B.5) can be similarly written using the generalized Lagrangian function

$$L = \sqrt{g_{\alpha\beta}\dot{x}_\alpha \dot{x}_\beta} \tag{B.8a}$$

where the action is

$$S = \int_{\lambda_0}^{\lambda} L(x_\alpha, \dot{x}_\alpha, \lambda)\, d\lambda \tag{B.8b}$$

Here the path is parametrized by λ, in which $\dot{x}_\alpha(\lambda) = \frac{dx_\alpha(\lambda)}{d\lambda}$ (see Fig. B.2).

An Euler–Lagrange equation can be written to minimize the path length as given in Appendix A.1.1:

$$\frac{\partial L}{\partial x_\mu} - \frac{d}{d\lambda}\frac{\partial L}{\partial \dot{x}_\mu} = 0 \tag{B.9}$$

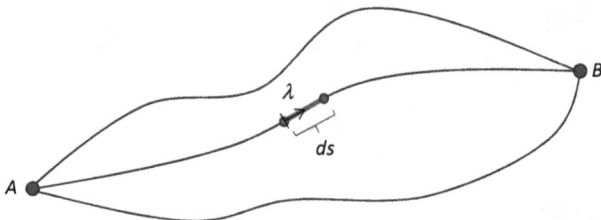

Figure B.2. Various paths between two fixed points A and B.

Using this Lagrangian formulation, the equation of a geodesic is [Lewis et al., 2022]:

$$\boxed{\ddot{x}_\nu + \dot{x}_\alpha \dot{x}_\delta \Gamma^\nu_{\delta\alpha} = 0}$$

(B.10)

where the Christoffel symbol is defined as

$$\Gamma^\nu_{\delta\alpha} = \frac{1}{2} g^{\mu\nu} [g_{\mu\alpha/\delta} + g_{\delta\mu/\alpha} - g_{\alpha\delta/\mu}]$$

(B.11)

and $g_{ij/k} \equiv \frac{\partial g_{ij}}{\partial x_k}$.

For example, for ordinary three-dimensional (Euclidean) flat space, $\nu = 1, 2, 3$ with coordinates x_1, x_2 and x_3. Therefore, for a Cartesian coordinate system, $g_{11} = g_{22} = g_{33} = 1$ and $g_{\mu\nu} = 0$ for $\mu \neq \nu$. Hence, $\Gamma^\nu_{\delta\alpha} = 0$ since the derivatives of the g's of the metric in Eq. (B.11) are zero. Thus, Eq. (B.10) reduces to $\ddot{x}_\nu = \frac{d^2 x_\nu}{d\lambda^2} = 0$, or in component form: $\frac{d^2 x_1}{d\lambda^2} = 0, \frac{d^2 x_2}{d\lambda^2} = 0, \frac{d^2 x_3}{d\lambda^2} = 0$. The solution of each component equation is a straight line for the geodesic where, with $x \equiv x_1, y \equiv x_2, z \equiv x_3$, one obtains: $x = x_o + \lambda a, y = y_o + \lambda b, z = z_o + \lambda c$. All other quantities are integration constants. Thus, solving for λ yields the Cartesian equation of a line: $\frac{x-x_o}{a} = \frac{y-y_o}{b} = \frac{z-z_o}{c}$.

An "Einstein–Hilbert action" can be employed using a scalar Lagrangian density L. For general relativity, this quantity is integrated over the four-volume of space–time V:

$$S = \int_V L d^4 V$$

(B.12)

To guarantee that the action S remains "invariant" so that there is no change using a different coordinate system, one takes $d^4 V = \sqrt{-g} d^4 dx$ where g is the determinant of the 4×4 metric $g_{\alpha\beta}$ as applicable to four-dimensional space–time. This quantity accounts for any type of surface and for any general coordinate system and arises with the general transformation of a coordinate system [Einstein, 2014]. The Lagrangian L needs to be a scalar derivable from the metric with no higher than second-order derivatives. This requirement is met by the Ricci scalar R (that is, $L = R$). The Ricci scalar is derived as a contraction of the Ricci tensor: $R = g^{\mu\nu} R_{\mu\nu}$, where the Ricci tensor physically accounts for the deviation of the metric from flatness. Thus, Eq. (B.12) becomes:

$$S = \int_V R\sqrt{-g} d^4 x = \int_V g^{\mu\nu} R_{\mu\nu} \sqrt{-g} d^4 x$$

(B.13)

The principle of least action requires $\delta S = 0$. The "variance" of the Einstein–Hilbert action in Eq. (B.13) is calculated as:

$$\delta S = \int_V [\delta g^{\mu\nu} R_{\mu\nu} \sqrt{-g} + g^{\mu\nu} \delta R_{\mu\nu} \sqrt{-g} + g^{\mu\nu} R_{\mu\nu} \delta \sqrt{-g}] d^4 x$$

$$= \int_V [\delta g^{\mu\nu} R_{\mu\nu} \sqrt{-g} + \delta R_{\mu\nu} g^{\mu\nu} \sqrt{-g} + \delta \sqrt{-g} R] d^4 x \qquad (B.14)$$

Variation of the Metric Determinant: The variation of $\sqrt{-g}$ in the last term of Eq. (B.14) can be obtained following the analysis of [Adler et al., 1975]. The determinant $g = |g_{\mu\nu}|$ can be evaluated with a Laplace expansion from linear algebra theory for the elements of the νth column and cofactors $\Delta^{\mu\nu}$:

$$g = \sum_\mu g_{\mu\nu} \Delta^{\mu\nu} \qquad (B.15)$$

Thus,

$$\frac{\partial g}{\partial g_{\mu\nu}} = \Delta^{\mu\nu} \Rightarrow \delta g = \frac{\partial g}{\partial g_{\mu\nu}} \delta g_{\mu\nu} = \Delta^{\mu\nu} \delta g_{\mu\nu} \qquad (B.16)$$

Applying the inverse of the matrix for the symmetrical matrix $g_{\mu\nu}$ gives [Adler et al., 1975]

$$g^{\mu\nu} = \frac{1}{g} \Delta^{\mu\nu} \Rightarrow \Delta^{\mu\nu} = g g^{\mu\nu} \qquad (B.17)$$

Inserting Eq. (B.17) into Eq. (B.16) yields

$$\delta g = g \left(g^{\mu\nu} \delta g_{\mu\nu} \right) \qquad (B.18)$$

Now given that $g^{\mu\nu} g_{\mu\nu} = \delta^\nu_\nu$ (where the general Kronecker delta function δ^ν_μ has constant values equal to unity when $\mu = \nu$ or is zero otherwise), therefore

$$\delta(g^{\mu\nu} g_{\mu\nu}) = g_{\mu\nu} \delta g^{\mu\nu} + g^{\mu\nu} \delta g_{\mu\nu} = 0 \Rightarrow g^{\mu\nu} \delta g_{\mu\nu} = -g_{\mu\nu} \delta g^{\mu\nu} \qquad (B.19)$$

Inserting Eq. (B.19) into Eq. (B.18) gives:

$$\delta g = -g g_{\mu\nu} \delta g^{\mu\nu} \qquad (B.20)$$

By differentiation

$$\delta \sqrt{-g} = -\frac{1}{2} \frac{1}{\sqrt{-g}} \delta g \qquad (B.21)$$

Substituting Eq. (B.20) into Eq. (B.21) yields

$$\boxed{\delta\sqrt{-g} = -\frac{1}{2}\sqrt{-g}\,g_{\mu\nu}\,\delta g^{\mu\nu}}$$

(B.22)

This expression can be subsequently substituted into Eq. (B.14) to give:

$$\delta S = \int_V \left[\delta g^{\mu\nu}R_{\mu\nu}\sqrt{-g} - \frac{1}{2}\sqrt{-g}\,g_{\mu\nu}\delta g^{\mu\nu}R + \delta R_{\mu\nu}g^{\mu\nu}\sqrt{-g}\right]d^4x$$

$$= \int_V \left(R_{\mu\nu} - \frac{1}{2}g_{\mu\nu}R\right)\delta g^{\mu\nu}\sqrt{-g}\,d^4x + \int_V g^{\mu\nu}\delta R_{\mu\nu}\sqrt{-g}\,d^4x \quad \text{(B.23)}$$

Vanishing Term: The last term $\int_V g^{\mu\nu}\delta R_{\mu\nu}\sqrt{-g}\,d^4x$ in Eq. (B.23) vanishes, where using the "Palatini identity" (see [Weinberg, 1972] and [d'Inverno, 1992]):

$$\delta R_{\mu\nu} = \nabla_\alpha(\delta^\alpha_{\mu\nu}) - \nabla_\nu(\delta^\alpha_{\mu\alpha})$$

(B.24)

so that:

$$\int_V g^{\mu\nu}\delta R_{\mu\nu}\sqrt{-g}\,d^4x = \int_V g^{\mu\nu}\left[\nabla_\alpha(\delta^\alpha_{\mu\nu}) - \nabla_\nu(\delta^\alpha_{\mu\alpha})\right]\sqrt{-g}\,d^4x$$

$$= \int_V \left[\nabla_\alpha(g^{\mu\nu}(\delta^\alpha_{\mu\nu})) - \nabla_\nu(g^{\mu\nu}(\delta^\alpha_{\mu\alpha}))\right]\sqrt{-g}\,d^4x$$

$$= \int_V \left[\nabla_\alpha(g^{\mu\nu}(\delta^\alpha_{\mu\nu})) - \nabla_\alpha(g^{\mu\alpha}(\delta^\alpha_{\mu\alpha}))\right]\sqrt{-g}\,d^4x$$

$$= \int_V \nabla_\alpha\left[g^{\mu\nu}\delta^\alpha_{\mu\nu} - g^{\mu\alpha}\delta^\alpha_{\mu\alpha}\right]\sqrt{-g}\,d^4x \quad \text{(B.25)}$$

The expression in square brackets in Eq. (B.25) is a rank 1 tensor: $A^\alpha \equiv g^{\mu\nu}\delta^\alpha_{\mu\nu} - g^{\mu\alpha}\delta^\alpha_{\mu\alpha}$, so that

$$\int_V g^{\mu\nu}\delta R_{\mu\nu}\sqrt{-g}\,d^4x = \int_V \nabla_\alpha A^\alpha\sqrt{-g}\,d^4x$$

(B.26)

Using the divergence theorem in vector calculus:

$$\int_V \nabla_\alpha A^\alpha dV = \int_S A^\alpha dS$$

(B.27)

the volume integral in Eq. (B.26) can be converted into a surface integral. However, the surface integral vanishes since variations are assumed to vanish on the surface of V so that

$$\int_V g^{\mu\nu}\delta R_{\mu\nu}\sqrt{-g}d^4x = 0 \tag{B.28}$$

Thus, inserting Eq. (B.28) into Eq. (B.23) and setting $\delta S = 0$ gives:

$$\delta S = \int_V \left(R_{\mu\nu} - \frac{1}{2}g_{\mu\nu}R \right)\delta g^{\mu\nu}\sqrt{-g}d^4x = 0 \tag{B.29}$$

Since $\delta g^{\mu\nu}$ is arbitrary, the quantity inside the parenthesis must equal zero and one therefore obtains the field equations for general relativity in the absence of matter or energy [Einstein, 2014]:

$$R_{\mu\nu} - \frac{1}{2}g_{\mu\nu}R = 0 \tag{B.30}$$

The Ricci tensor $R_{\mu\nu}$ in Eq. (B.30) is obtained by contracting the Riemann curvature tensor $R^\alpha_{\mu\sigma\nu}$:

$$R^\alpha_{\mu\sigma\nu} = \frac{\partial\Gamma^\alpha_{\mu\nu}}{\partial x_\sigma} - \frac{\partial\Gamma^\alpha_{\mu\sigma}}{\partial x_\nu} + \Gamma^\alpha_{\sigma\gamma}\Gamma^\gamma_{\mu\nu} - \Gamma^\alpha_{\nu\gamma}\Gamma^\gamma_{\mu\sigma} \tag{B.31}$$

where the Christoffel symbol, as defined in Eq. (B.11) and used in differential geometry, describes the changes in basis vectors in a coordinate system. It physically relates to fictitious forces as induced by a non-inertial (i.e., accelerating) reference frame. Hence, contracting Eq. (B.31) on the first and third indices (with $\sigma = \alpha$), one obtains $R_{\mu\nu}$ as needed in Eq. (B.30):

$$R_{\mu\nu} = R^\alpha_{\mu\alpha\nu} = \frac{\partial\Gamma^\alpha_{\mu\nu}}{\partial x_\alpha} - \frac{\partial\Gamma^\alpha_{\mu\alpha}}{\partial x_\nu} + \Gamma^\alpha_{\alpha\gamma}\Gamma^\gamma_{\mu\nu} - \Gamma^\alpha_{\nu\gamma}\Gamma^\gamma_{\mu\alpha} \tag{B.32}$$

There is symmetry in the lower indices of the Christoffel symbol and one can interchange symbols for repeated (dummy) indices for comparison of the Riemann curvature tensor and Ricci tensor presented in other treatise.

B.2.1 *General field equations*

The action in Eq. (B.13) for the vacuum equation can be further generalized to account for both matter and energy as well as a "cosmological constant". This added constant is currently thought to explain the occurrence of "dark energy", which is believed to be responsible for the observed expansion of the universe [Dodelson, 2003]. Thus, now including both a source Lagrangian, L_m, and cosmological constant, Λ:

$$S = \int_V \left[\frac{1}{2\kappa} (R - 2\Lambda) + L_m\right] \sqrt{-g} d^4x \tag{B.33}$$

The coupling constant κ can be determined by linearizing the Einstein field equations. Thus, κ can be evaluated by matching this linearized field equation for a slowly varying weak gravitational field for bodies with small velocities to Newton's classical law of gravitation.

Similarly, for the generalized field equations, one can apply a variation of the action:

$$\delta S = \delta \int_V [(R - 2\Lambda) + 2\kappa L_m] \sqrt{-g} d^4x = 0 \tag{B.34}$$

Taking the variation within the integral sign of Eq. (B.34), using the product rule and the definition of the Ricci scalar gives:

$$\delta S = \underbrace{\int_V \delta \left[g^{\mu\nu} R_{\mu\nu} \sqrt{-g}\right] dx^4}_{eq.\ (B.29)} - 2\Lambda \int_V \underbrace{\delta \left[\sqrt{-g}\right]}_{eq.\ (B.22)} dx^4$$

$$+ 2\kappa \int_V \left\{ \delta [L_m] \sqrt{-g} + L_m \underbrace{\delta \left[\sqrt{-g}\right]}_{eq.\ (B.22)} \right\} d^4x = 0 \tag{B.35}$$

Replacing the first term of Eq. (B.35) with Eq. (B.29), as well as the expression for $\delta\sqrt{-g}$ using Eq. (B.22), Eq. (B.35) becomes:

$$\delta S = \int_V \left(R_{\mu\nu} - \frac{1}{2} g_{\mu\nu} R\right) \delta g^{\mu\nu} \sqrt{-g} d^4x + \int_V (\Lambda g_{\mu\nu}) \delta g^{\mu\nu} \sqrt{-g} d^4x$$

$$+ \int_V \kappa \left\{ 2 \frac{\delta L_m}{\delta g^{\mu\nu}} \delta g^{\mu\nu} \sqrt{-g} - (L_m\, g_{\mu\nu}) \delta g^{\mu\nu} \sqrt{-g} \right\} d^4x = 0 \tag{B.36}$$

Thus, collecting terms in Eq. (B.36):

$$\delta S$$

$$= \int_V \left\{ \left(R_{\mu\nu} - \frac{1}{2} g_{\mu\nu} R\right) + (\Lambda g_{\mu\nu}) + \kappa \underbrace{\left[2\frac{\delta L_m}{\delta g^{\mu\nu}} - (L_m\, g_{\mu\nu})\right]}_{= -T_{\mu\nu}} \right\} \delta g^{\mu\nu} \sqrt{-g}\, d^4x$$

$$= \int_V \left\{ \left(R_{\mu\nu} - \frac{1}{2} g_{\mu\nu} R\right) + (\Lambda g_{\mu\nu}) - (\kappa T_{\mu\nu}) \right\} \delta g^{\mu\nu} \sqrt{-g}\, d^4x = 0 \tag{B.37}$$

Here, the stress-energy tensor $T_{\mu\nu}$ is defined as:

$$T_{\mu\nu} = -\left[2\frac{\delta L_m}{\delta g^{\mu\nu}} - L_m\, g_{\mu\nu}\right] \qquad (B.38)$$

Again, since $\delta g^{\mu\nu}$ is arbitrary for the integral, the integrand must equal zero yielding the complete field equations [d'Inverno, 1992]:

$$R_{\mu\nu} - \frac{1}{2}Rg_{\mu\nu} + \Lambda g_{\mu\nu} - \kappa T_{\mu\nu} = 0 \qquad (B.39a)$$

or

$$\boxed{G_{\mu\nu} + \Lambda g_{\mu\nu} = \frac{8\pi G}{c^4}T_{\mu\nu}} \qquad (B.39b)$$

where $G_{\mu\nu} = R_{\mu\nu} - \frac{1}{2}Rg_{\mu\nu}$ is the Einstein tensor and the coupling constant $\kappa = \frac{8\pi G}{c^4}$ as follows in the Newtonian limit. Here G is the gravitational constant and c is the speed of light. As expected, without the presence of matter or energy (that is, where $\Lambda = \kappa = 0$), Eq. (B.39a) reduces to Eq. (B.30).

B.2.2 *Newtonian limit*

Consider a slowly-varying weak gravitational field, where the metric only differs slightly from flat space such that

$$g_{ab} = \eta_{ab} + \epsilon h_{ab} \qquad (B.40)$$

With flat space, the Minkowski metric is given by

$$\eta_{ab} = \begin{bmatrix} 1 & 0 & 0 & 0 \\ 0 & -1 & 0 & 0 \\ 0 & 0 & -1 & 0 \\ 0 & 0 & 0 & -1 \end{bmatrix}$$

The small dimensionless parameter ϵ is of the order of (v/c) for a body with a velocity v much less than the speed of light c.

Assuming a coordinate system $(x^a) = (x^0, x^1, x^2, x^3) = (x^0, x^\alpha) = (ct, x, y, z)$, with the lowest order for ϵ: $\delta x^\alpha = (v/c)c\delta t \sim \epsilon \delta x^0$, for any function f: $\frac{\partial f}{\partial x^0} \sim \epsilon\frac{\partial f}{\partial x^\alpha}$. Consider the time-like geodesic equation as parametrized by a proper time τ [d'Inverno, 1992]:

$$\frac{d^2 x^a}{d\tau^2} + \Gamma^a_{bc}\frac{dx^b}{d\tau}\frac{dx^c}{d\tau} = 0 \qquad (B.41)$$

where, $c^2 d\tau^2 = ds^2 = c^2 dt^2 - dx^2 - dy^2 - dz^2 = dt^2(c^2 - v^2) = c^2 dt^2 (1 - \epsilon^2)$.
Thus, to first order $d\tau \sim dt$. Using the spatial part of the geodesic equation
(that is, $a = \alpha$ and $b = c = 0$, along with $x^0 = ct$), this equation reduces in
the slow-motion approximation to:

$$\frac{1}{c^2}\frac{d^2 x^\alpha}{dt^2} + \frac{1}{c^2}\Gamma^\alpha_{bc}\frac{dx^b}{dt}\frac{dx^c}{dt} = 0 \Rightarrow \frac{1}{c^2}\frac{d^2 x^\alpha}{dt^2} + \Gamma^\alpha_{00} = 0. \tag{B.42}$$

From the definition of the Christoffel symbol:

$$\Gamma^a_{bc} = \frac{1}{2}g^{ad}[g_{bd/c} + g_{cd/b} - g_{bc/d}] \sim \frac{1}{2}\eta^{ad}\,\epsilon[h_{bd/c} + h_{cd/b} - h_{bc/d}] \tag{B.43}$$

Therefore, with $b = c = 0$ and $d \to \alpha$ with $\eta^{\alpha\alpha} = -1$ along the space part
of the diagonal of the Minkowski metric, the Christoffel symbol reduces to
$\Gamma^\alpha_{00} = -\frac{1}{2}\epsilon[2h_{0\alpha/0} - h_{00/\alpha}] = -\frac{1}{2}\epsilon\left[2\frac{\partial h_{0\alpha}}{\partial x^0} - \frac{\partial h_{00}}{\partial x^\alpha}\right] \sim \frac{1}{2}\epsilon\frac{\partial h_{00}}{\partial x^\alpha} + O(\epsilon^2)$. For this
latter result, the first term for Γ^α_{00} is of lower-order and can be neglected
since $\frac{\partial h_{00}}{\partial x^0} \sim \epsilon\frac{\partial h_{00}}{\partial x^\alpha}$. Applying this expression for Γ^α_{00} in Eq. (B.42) one
obtains:

$$\frac{d^2 x^\alpha}{dt^2} = -\frac{c^2}{2}\epsilon\frac{\partial h_{00}}{\partial x^\alpha} \tag{B.44}$$

On comparison of Eq. (B.44) with the Newtonian equation:

$$\frac{d^2 x^\alpha}{dt^2} = -\frac{\partial \phi}{\partial x^\alpha} \tag{B.45}$$

$\phi = \frac{c^2}{2}\epsilon h_{00}$. On solving for ϵh_{00} for the perturbed metric $g_{00} = \eta_{00} + \epsilon h_{00}$,
and noting that the diagonal element $\eta_{00} = 1$, the weak-field limit gives:

$$g_{00} \sim 1 + \frac{2\phi}{c^2} \tag{B.46}$$

The non-vacuum field equations containing the Ricci tensor and scalar in
Eq. (B.39a), when linearized for the slightly perturbed metric in Eq. (B.40)
reduce to [d'Inverno, 1992]:

$$\frac{1}{2}\epsilon\Box\left[h_{ab} - \frac{1}{2}\eta_{ab}\eta^{cd}h_{cd}\right] = -\kappa T_{ab} \tag{B.47}$$

where the d'Alembertian operator is $\Box = \frac{\partial^2}{\partial t^2} - \nabla^2$ and ∇^2 is the Laplacian
operator. For the slow-moving approximation, the time derivative can be
neglected so that Eq. (B.47) reduces to

$$\frac{\epsilon}{2}\nabla^2\left[h_{ab} - \frac{1}{2}\eta_{ab}\eta^{cd}h_{cd}\right] = \kappa T_{ab} \tag{B.48}$$

Contracting Eq. (B.48) with η^{ab} gives

$$\frac{\epsilon}{2}\nabla^2\left[\eta^{ab}h_{ab} - \frac{1}{2}\eta^{ab}\eta_{ab}\eta^{cd}h_{cd}\right] = \kappa\eta^{ab}T_{ab} \qquad (B.49)$$

However, $\eta^{ab}\eta_{ab} = \delta^a_a$, where the Kronecker delta function $\delta^a_a = 3$ on summing the repeated indices using the Einstein convention over the three spatial dimensions. Hence, with this result and replacing the indices a and b by c and d, respectively, since they are dummy indices, Eq. (B.49) becomes:

$$\frac{\epsilon}{2}\nabla^2\left[\eta^{cd}h_{cd} - \frac{3}{2}\eta^{cd}h_{cd}\right] = \kappa\eta^{cd}T_{cd} \Rightarrow -\frac{\epsilon}{2}\nabla^2\left[\eta^{cd}h_{cd}\right] = \kappa\eta^{cd}T_{cd} \quad (B.50)$$

Multiplying Eq. (B.50) through by $\frac{1}{2}\eta_{ab}$ to match the the second term in Eq. (B.48) gives

$$-\frac{\epsilon}{2}\nabla^2\left[\frac{1}{2}\eta_{ab}\eta^{cd}h_{cd}\right] = \frac{1}{2}\kappa\eta_{ab}\eta^{cd}T_{cd} \qquad (B.51)$$

Thus, substituting Eq. (B.51) into Eq. (B.48) gives

$$\frac{\epsilon}{2}\nabla^2 h_{ab} = \kappa T_{ab} - \frac{1}{2}\kappa\eta_{ab}\eta^{cd}T_{cd} \qquad (B.52)$$

Here the energy-momentum tensor $T_{ab} = c^2\rho_0\delta^0_a\delta^0_b$ where ρ_0 is the material density and c is the speed of light. Thus, $\eta_{cd}T_{cd} = c^2\rho_0$. Taking the zero-zero component:

$$\frac{\epsilon}{2}\nabla^2 h_{00} = \kappa T_{00} - \frac{1}{2}\kappa T_{00} \qquad (B.53)$$

so that

$$\epsilon\nabla^2 h_{00} = c^2\kappa\rho_0 \qquad (B.54)$$

Taking the Laplacian on both sides of Eq. (B.40), and using Eq. (B.54), one obtains $\nabla^2 g_{00} = \epsilon\nabla^2 h_{00} = c^2\kappa\rho_0$. Again, taking the Laplacian on both sides of Eq. (B.46), $\nabla^2 g_{00} = \frac{2}{c^2}\nabla^2\phi$. Equating these two expressions:

$$\nabla^2\phi = \frac{1}{2}c^4\kappa\rho_0 \qquad (B.55)$$

Furthermore, comparing Eq. (B.55) to Poisson's equation for Newtonian gravity:

$$\nabla^2\phi = 4\pi G\rho_0 \qquad (B.56)$$

the coupling constant is deduced as

$$\kappa = 8\pi G/c^4 \tag{B.57}$$

where G is the gravitational constant.

B.2.3 *Gravitational waves*

The linearized Einstein tensor for the slightly perturbed metric from flat space in Eq. (B.40) is given by [d'Inverno, 1992]:

$$G_{ab} = \frac{\epsilon}{2}\left[h^c_{a,bc} + h^c_{b,ac} - \Box h_{ab} - h_{,ab} - \eta_{ab}h^{cd}_{,cd} + \eta_{ab}\Box h\right] \tag{B.58}$$

where $h \equiv \eta^{cd}h_{cd} = h^c_c$, $h^{cd}_{,cd} \equiv \frac{\partial^2 h^{cd}}{\partial x^c \partial x^d}$ and $h_{,ab} \equiv \frac{\partial^2 h}{\partial x^a \partial x^b}$. The linearized equations can be further transformed using a so-called "gauge transformation", a methodology used in electrodynamics where the fields remain invariant under such transformations. This type of transformation importantly leaves R^a_{bcd}, R_{ab} and R unchanged. Thus, using the coordinate transformation:

$$x^a \rightarrow x'^a = x^a + \epsilon\xi^a \tag{B.59}$$

Eq. (B.59) gives rise to coordinate-transformation changes in h_{ab}:

$$h_{ab} \rightarrow h'_{ab} = h_{ab} - \xi_{a,b} - \xi_{b,a} \tag{B.60}$$

A new variable ψ_{ab} can be further defined as

$$\psi_{ab} \equiv h_{ab} - \frac{1}{2}\eta_{ab}h \tag{B.61}$$

This function is known as a "trace reverse" representation since it has the property $\psi = \psi^c_c = -h$. With this new variable, a so-called "de Donder gauge" or "Einstein gauge" can be imposed, where $\psi^a_{b,a} = 0$ (or equivalently $\psi_{ab} \rightarrow \psi'_{ab} = \psi_{ab} - \xi_{a,b} - \xi_{b,a} + \eta_{ab}\xi^c_{,c}$). As such, the full field equations for the linearized Einstein tensor with the given gauge (or equivalent coordinate) transformation can be written as (dropping the primes) [d'Inverno, 1992]:

$$\frac{1}{2}\epsilon\Box\psi_{ab} = -\kappa T_{ab} \tag{B.62}$$

or in vacuum:

$$\Box\psi_{ab} = 0 \tag{B.63}$$

Moreover, multiplying Eq. (B.63) by η^{ab} gives

$$\eta^{ab}\Box\psi_{ab} = \Box\left(\eta^{ab}\psi_{ab}\right) = \Box\psi^c_c = \Box\psi = -\Box h = 0 \tag{B.64}$$

Thus, using Eq. (B.61), Eq. (B.63) and Eq. (B.64), one obtains a wave equation, where "gravitational waves" are predicted to occur that travel at the speed of light according to

$$\Box h_{ab} = 0 \qquad (B.65)$$

B.2.4 *Unification of gravity and electromagnetism*

As shown in Appendix A.2, in the covariant formalism for Maxwell's equations, A_μ is a four-vector whose time component is the scalar potential ϕ and the spatial component is the vector potential $\vec{A}$. As mentioned, a "gauge transformation" can be further employed by shifting $A_\mu \to A_\mu + \partial_\mu \Lambda(x)$ where $\Lambda(x)$ is an arbitrary function of space–time, without changing Maxwell's equations that govern the dynamics of the electromagnetic field.

As described in Chapter 5 for the precursor of string theory, Kaluza extended the theory of general relativity in order to unify both gravity and electromagnetism with the idea of an extra fifth dimension. In this theory, one considers the Newtonian limit in Appendix B.2.2, where $g_{\mu\nu}$ deviates from the flat metric $\eta_{\mu\nu}$ by only a small amount. Furthermore, two sets of coordinates are assumed to differ by the small quantity: $\epsilon^\mu = x'^\mu(x) - x^\mu$. As such, the gravitational field $h_{\mu\nu}$ will transform as $h_{\mu\nu} \to h_{\mu\nu} + \partial_\mu \epsilon_\nu + \partial_\nu \epsilon_\mu$. This transformation, in fact, is similar in form to the electromagnetic transformation $A_\mu \to A_\mu + \partial_\mu \Lambda$. This resemblance suggests some kind of unification that led Kaluza to propose a 5th dimension for space–time. Here instead one has the index $M = (0, 1, 2, 3, 4, 5) = (\mu, 5)$. In this space–time representation, the gravitation field h_{MN} transforms as: $h_{MN} \to h_{MN} + \partial_M \epsilon_N + \partial_N \epsilon_M$. This implies that $h_{\mu\nu} \to h_{\mu\nu} + \partial_\mu \epsilon_\nu + \partial_\nu \epsilon_\mu$ and also that $h_{\mu5} \to h_{\mu5} + \partial_\mu \epsilon_5 + \partial_5 \epsilon_\mu$, which becomes $h_{\mu5} \to h_{\mu5} + \partial_\mu \epsilon_5$ if ϵ_μ does not depend on x. Now, simply renaming $h_{\mu5} = A_\mu$ and $\epsilon_5 = \Lambda$, one recovers $A_\mu \to A_\mu + \partial_\mu \Lambda$ and hence electromagnetism.

B.3 Experimental Tests of General Relativity Theory

Relativity is a "falsifiable theory" in that it can be tested by experiment. Three classical experimental tests of general relativity include the prediction of

(i)   The advance of the perihelion of the planet of Mercury;
(ii) The bending of light in a gravitational field;
(iii) A gravitational red-shift.

B.3.1   *Advance of the perihelion of mercury*

In accordance with the principle of equivalence, an object is not constrained by any external force even while moving in the presence of matter. This includes the orbit of a planet around the sun. As shown in Newtonian physics, a planet will trace out an ellipse with the central force of gravity from the sun located at a foci. The ellipse is not fixed where a slight shift is observed each time the planet orbits. The shift in the perihelion, which is the point of nearest approach to the sun, is more pronounced for the planet Mercury. This shift can be calculated more precisely by general relativity and compared to a measurement of 0.001 degrees per century.

The path of a planet follows a geodesic in the general theory of relativity. This timelike geodesic can be obtained from the so-called "Schwarzschild solution" of the general field equations. The line element for the Schwarzschild solution follows from a generalization of the Minkowski metric, which is recast in spherical coordinates (t, r, θ, ϕ) instead of Cartesian ones (t, x, y, z). As such, it can be shown that:

$$ds^2 = \left(1 - \frac{2m}{r}\right)c^2 dt^2 - \frac{dr^2}{\left(1 - \frac{2m}{r}\right)} - r^2\left(d\theta^2 + sin^2\theta d\phi^2\right) \tag{B.66}$$

where m is the "geometric mass", defined as $m = GM/c^2$ where M is the mass of the sun. For large radial distances, as $r \to 0$, this metric approaches that for flat space.

A geodesic can be computed for a given metric as described using the methodology in Appendix B.2. This approach makes use of a Lagrangian functional to describe the equations of motion of the system (i.e., Eq. (B.8a) and Eq. (B.8b)) leading to the development of a geodesic equation (i.e., Eq. (B.10)). Applying this approach for a time-like geodesic with planetary motion in an equatorial $(x - y)$ plane $(\theta = \pi/2)$, yields the equations [d'Inverno, 1992], [Lieber and Lieber, 1966], [McMahon, 2006]:

$$\frac{d^2u}{d\phi^2} + u = \frac{m}{h^2} + 3mu^2 \tag{B.67a}$$

and

$$r^2 \frac{d\phi}{ds} = h \tag{B.67b}$$

where h is a constant and $u = 1/r$. The orbital differential equation in Eq. (B.67a) is in fact the relativistic version of "Binet's equation". In particular, Eq. (B.67a) differs from the Newtonian result with the addition of the last term. This extra term provides a more accurate prediction of the advance of the perihelion of Mercury thereby providing an important validation of general relativity.

One can solve Eq. (B.67a) with a perturbation method. Consider a solution of constant radius $u_c = 1/r_c$. In this case, Eq. (B.67a) gives

$$\overbrace{\cancel{\frac{d^2 u_c}{d\phi^2}}}^{0} + u_c = \frac{m}{h^2} + 3mu_c^2 \tag{B.68}$$

A solution of the form can be assumed:

$$u(\phi) = u_c[1 + v(\phi)] \tag{B.69}$$

Inserting Eq. (B.69) into Eq. (B.67a), using the relation in Eq. (B.68) and dividing the resultant equation through by u_c, gives

$$\frac{d^2 v}{d\phi^2} + v\left(1 - \frac{6m}{r_c}\right) \approx 0 \tag{B.70}$$

Here a term of the order of v^2 has been neglected since $v \ll 1$. This equation has the form of a harmonic oscillator, with the solution:

$$v(\phi) = A\cos(\omega\phi + \phi_0) \tag{B.71}$$

where the $\omega = \sqrt{1 - \frac{6m}{r_c}}$. This latter expression corresponds to the period:

$$T = \frac{2\pi}{\sqrt{1 - \frac{6m}{r_c}}} \sim 2\pi(1 + \epsilon) \tag{B.72}$$

where $\epsilon = \dfrac{3m}{r_c}$ since $r_c \gg m$. In non-relativistic units, $m = GM/c^2$. Hence, the orbit is still periodic but no longer of period 2π.

For the derivation of the advance of the perihelion, one can use the quantity of the semi-latus rectum for Kepler motion in an ellipse, where the parameter $r_c = a(1 - e^2)$ [d'Inverno, 1992]. Here a is the semi-major axis and e is the eccentricity of the orbit. Kepler's third law for the period of the orbit is: $T^2 = \left(\frac{4\pi^2}{GM}\right)a^3$. Thus, an expression for the advance of the perihelion in Eq. (B.72) from general relativity is derived as

$$2\pi\epsilon = 2\pi \left[\frac{3GM/c^2}{(1-e^2)\,a} \right] \left[\frac{(4\pi^2/GM)\,a^3}{T^2} \right] = \frac{24\pi^3 a^2}{c^2 T^2 (1-e^2)} \qquad \text{(B.73)}$$

This expression was proposed and evaluated by Einstein as a classic test of his theory [d'Inverno, 1992]. The result in Eq. (B.73) can be evaluated using the speed of light as $c = 3.00 \times 10^8$ m s^{-1}. For Mercury, the semi-major axis is $a = 5.79 \times 10^{10}$ m, the eccentricity $e = 0.206$ and the period $T = 87.97$ d or 7.6×10^6 s. Thus, the apsidal precession during one period of revolution is 5.01×10^{-7} rad $\times (180/\pi)$ deg/rad $\times$ 3600 arcsecond/deg $= 0.103$ arcsecond. In one hundred years, Mercury makes $(365.25 \text{ d/y} \times 100 \text{ y})/(87.97 \text{ d/revolution}) = 415$ revolutions around the sun. Therefore in a century, the precession is 0.103 arcsecond per revolution $\times$ 415 revolutions/century $= 43$ arcsecond per century, which matches the measured value.

B.3.2   *Bending of light*

General relativity also predicts that the path of a ray of light will bend as it passes near a large mass such as the sun. The presence of matter makes the space non-Euclidean so that the path of any freely moving object (be it a planet or a light beam) will move along a geodesic. Thus, light coming from a distant star is predicted to be bent toward the sun, which can be checked with an astronomical expedition during a solar eclipse. The position of a star can be checked with its location in the sky when the sun is not present (see Fig. B.3).

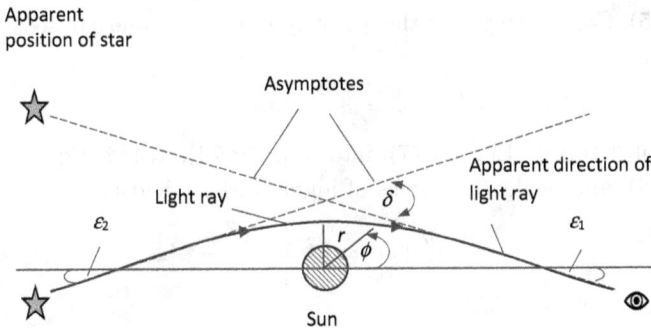

Figure B.3. Deflection of light in a gravitational field.

A ray of light in vacuum travels on a "null geodesic" in space–time in accordance with special relativity. For Euclidean space–time, $ds^2 = c^2 dt^2 - (dx^2 + dy^2 + dz^2)$. Dividing this expression through by dt^2 gives

$$\frac{ds^2}{dt^2} = c^2 - \left(\frac{dx^2}{dt^2} + \frac{dy^2}{dt^2} + \frac{dz^2}{dt^2} \right)$$

Here $dx/dt, dy/dt$ and dz/dt are components of the velocity, v, so that

$$\frac{ds^2}{dt^2} = c^2 - v^2.$$

However, with light travelling on a null geodesic, $ds^2 = 0$, which yields $v = c$. Hence, the speed of light is simply c. For a non-Euclidean world, light also moves on a null geodesic. Hence, from Eq. (B.67b), $h \to \infty$ when $ds = 0$ so that Eq. (B.67a) simplifies to

$$\boxed{\frac{d^2 u}{d\phi^2} + u = 3mu^2} \tag{B.74}$$

Following the derivation of [d'Inverno, 1992], in the limit of special relativity m vanishes so that

$$\frac{d^2 u_0}{d\phi^2} + u_0 = 0 \tag{B.75}$$

This equation has the solution

$$u_0 = \frac{1}{D} \sin(\phi - \phi_0) \tag{B.76}$$

where D is a constant which corresponds to a distance of closest approach and one can take $\phi_0 = 0$ for convenience for the other integration constant. Again, one can consider that the equation of a light ray in Schwarzchild space–time in Eq. (B.74) is simply a perturbation of its behaviour in Eq. (B.75). Considering that the quantity m is small, one can seek solutions of the form:

$$u = u_0 + 3mu_1 \tag{B.77}$$

Thus, substituting Eq. (B.77) into Eq. (B.74), using Eq. (B.75) and Eq. (B.76), and neglecting terms of higher order, yields:

$$u_1'' + u_1 = u_0^2 = \frac{\sin^2 \phi}{D^2} \tag{B.78}$$

The solution of Eq. (B.78) is

$$u_1 = \frac{\left(1 + C \cos \phi + \cos^2 \phi \right)}{3D^2} \tag{B.79}$$

where C is a constant of integration. Thus, using Eq. (B.76), Eq. (B.77) and Eq. (B.79) yield the general solution

$$u \simeq \frac{1}{D} \sin \phi + \frac{m(1 + C \cos \phi + \cos^2 \phi)}{D^2} \qquad (B.80)$$

The angle of the asymptotes in Fig. B.3 are $-\epsilon_1$ and $\pi + \epsilon_2$. With these small angles substituted in for the angle ϕ in Eq. (B.80), one obtains the two small-angle expressions:

$$-\frac{\epsilon_1}{D} + \frac{m}{D^2}(2 + C) \quad \text{and} \quad -\frac{\epsilon_2}{D} + \frac{m}{D^2}(2 - C) \qquad (B.81)$$

Hence, the angle of defection $\delta = \epsilon_1 + \epsilon_2$ is obtained by adding the two equations together in Eq. (B.81) to yield:

$$\delta = \frac{4m}{D} \implies \delta = \frac{4GM}{c^2 D} \qquad (B.82)$$

where the latter expression is in non-relativistic units. Inserting in the mass of the sun, $M = 1.99 \times 10^{30}$ kg, the radius of the sun for a light ray that just grazes the sun, $D = 6.96 \times 10^8$ m, along with the gravitational constant, $G = 6.67 \times 10^{-11}$ m^3 kg^{-1} s^{-2}, and speed of light, $c = 3.00 \times 10^8$ m s^{-1}, yields a value of $\delta = 8.48 \times 10^{-6}$ rads $\times (180/\pi)$ deg/rad $\times$ 3600 arcsecond/deg $= 1.75$ seconds of arc. This predicted value is in reasonable agreement with early measurements that ranged from 1.57 to 1.82 ± 0.2 seconds of arc.

B.3.3   *Gravitational red-shift*

A gravitational red-shift will result as a consequence of the equivalence principle along with a "Doppler effect" in a gravitational field. For instance, from the principle of equivalence, gravitational effects at the surface of the earth are locally indistinguishable from that of a free-falling observer in space with an acceleration of g. Thus, in a laboratory experiment where a light pulse is emitted at the floor of a laboratory, a free-falling observer will say that by the time the signal has reached the ceiling, the ceiling has accelerated away from the light pulse. Therefore, when observed by a detector fixed to the ceiling, the signal will have been red-shifted. As such, the free-falling observer sees a kinematic Doppler shift, while the laboratory observer sees a gravitational red-shift.

In particular, in a gravitational field, a loss of energy results in a decrease in the wave frequency and a longer wavelength, corresponding with a shift towards the red end of the electromagnetic spectrum. A gravitational red-shift can also be interpreted as a result of a gravitational time dilation where an oscillator at a higher gravitational potential, which is higher from an attracting body, will tick faster from that observed at the same location as the oscillator at the lower gravitational potential. A red-shift also arises from a relativistic "Doppler effect" caused by the relative motion between a source and observer taking into consideration the effect of time dilation in special relativity on the observed frequencies.

As an analytic derivation of a gravitational red-shift, consider two observers carrying clocks on spatial world-lines for the paths that are traced out in the four-dimensional space–time for the coordinate system $(x^a) = (x^0, x^\alpha)$ (see Fig. B.4). Here x^0 is the world time and x^α is the spatial coordinate. The two world lines are x_1^α and x_2^α, respectively. The first observer sends out a signal to the second observer. The time of separation between successive wave crests as measured by the first clock in terms of a proper time (i.e., the time measured by a clock that follows the world-line) is $d\tau$, and by dx_1^0 in terms of the coordinate time. Thus, as shown by [d'Inverno, 1992], $d\tau^2 = g_{00}(dx_1^a)dx_1^0$. The corresponding interval recorded by the second observer is $\alpha d\tau$ in proper time $d\tau$ and coordinate time dx_2^0. Here α records how many times the second clock has ticked between the reception of two wave crests. Similarly, $d\tau^2 = g_{00}(dx_2^a)dx_2^0$. For the simplifying case of a static space–time, where $dx_1^0 = dx_2^0$ in Fig. B.4:

$$\alpha = \left[\frac{g_{00}(x_2^\alpha)}{g_{00}(x_1^\alpha)} \right]^{1/2} \tag{B.83}$$

It follows that if the atomic system has a characteristic frequency ν_0, then the second observer will measure a frequency for the first clock of

$$\bar{\nu}_0 = \frac{\nu_0}{\alpha} = \nu_0 \left[\frac{g_{00}(x_2^\alpha)}{g_{00}(x_1^\alpha)} \right]^{1/2} \tag{B.84}$$

Defining the frequency shift z:

$$z = \frac{\bar{\nu}_0 - \nu_0}{\nu_0} = \frac{\bar{\nu}_0}{\nu_0} - 1 \tag{B.85}$$

Thus using Eq. (B.84) and the weak field limit in Eq. (B.46),

$$\frac{\bar{\nu}_0}{\nu_0} = \frac{\left(1 + \frac{2\phi_1}{c^2} \right)^{1/2}}{\left(1 + \frac{2\phi_2}{c^2} \right)^{1/2}} \tag{B.86}$$

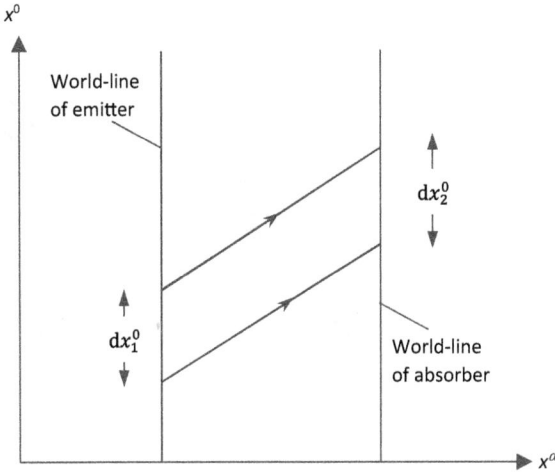

Figure B.4. Emission and reception of successive wave crests of a signal in static space–time.

Further, employing a binomial approximation since $2\phi/c^2 \ll 1$, the expression $\frac{1}{(1+2\phi/c^2)^{1/2}} \sim 1 + \phi/c^2$. Hence, using this approximation, Eq. (B.85) becomes:

$$z = \frac{1}{c^2}\left(\phi_2 - \phi_1\right) \tag{B.87}$$

Furthermore, substituting in the well known gravitational potential in Eq. (B.87)

$$\phi = -\frac{GM}{r} \tag{B.88}$$

one obtains the frequency shift

$$\phi = -\frac{GM}{c^2}\left[\frac{1}{r_1} - \frac{1}{r_2}\right] \tag{B.89}$$

Since the radial distance $r_1 < r_2$, therefore $z < 0$ and the frequency is red-shifted.

C Quantum Theory

Matter fundamentally behaves in accordance with quantum theory. "Quantum mechanics" (Appendix C.1) accounts for the physical properties of atoms and subatomic particles as indestructible particles. "Quantum field theory" covers a broader theoretical framework, which combines classical field theory, special relativity and quantum mechanics, to describe the behaviour of subatomic particles and their interaction (including annihilation and creation) (Appendix C.2). "Loop quantum gravity" (Appendix C.3) provides an approach to describe quantum gravity with a merging of quantum mechanics and the general theory of relativity.

C.1 Quantum Mechanics

Information about a particle is contained by the state of the particle or system that is represented by a "wave function" $\psi(x,t)$. This function assigns a complex number to each point x at each time t. The wave function can be thought of as a probability amplitude. This wave function, for example, is a solution of the following position-space "Schrödinger equation" for a non-relativistic particle of mass m:

$$i\hbar\frac{\partial\psi(x,t)}{\partial t} = \left[-\frac{\hbar^2}{2m}\frac{\partial^2}{\partial x^2} + V(x,t)\right]\psi(x,t) \qquad (C.1)$$

The particle is moving in a constrained environment of potential V, $\hbar$ is the reduced Planck's constant (i.e., $\hbar = h/2\pi$), and the constant i is the imaginary unit where $i = \sqrt{-1}$.

The position-space wave function ψ in Eq. (C.1) can be written more generally as the inner vector product of a time-dependent state vector $|\psi(t)\rangle$

with the "position eigenstates" $|x\rangle$:

$$\psi(x,t) = \langle x|\psi(t)\rangle \tag{C.2}$$

The form of the Schrödinger equation depends on the physical situation, with the most general form for a system evolving with time:

$$i\hbar\frac{d}{dt}|\psi(t)\rangle = \hat{H}|\psi(t)\rangle \tag{C.3}$$

where $|\psi(t)\rangle$ is the state vector of the quantum system and $\hat{H}$ is an observable called the "Hamiltonian operator". This operator corresponds to the total energy of the system, including both kinetic and potential energy. The set of energy eigenvalues corresponds to the possible outcomes from a measurement of the system's total energy. This equation provides a general framework in quantum theory by employing an appropriate expression for the Hamiltonian. This framework includes the further derivation of the Dirac equation to incorporate relativity theory.

For application of the Schrödinger equation, one writes down the Hamiltonian for the system that details the kinetic and potential energies of the system. This resulting partial differential equation for the wave function is solved, providing information about the system. The square of the absolute value of the wave function at each point defines a "probability density function" $|\psi(x,t)|^2$, which gives the probability that the particle or system will be found in a given state.

If the Hamiltonian is not dependent on time explicitly, stationary states can be solved by considering the time-independent Schrödinger equation:

$$\hat{H}|\psi\rangle = E|\psi\rangle \tag{C.4}$$

where E is the energy of the system. In accordance with linear algebra theory, Eq. (C.4) is an eigenvalue problem, where the wave function is the eigenfunction of the Hamiltonian operator with eigenvalue E [Lewis *et al.*, 2022]. Some wave functions may be independent of time resulting in a "static wave function". For example, a single electron in an unexcited hydrogen atom can be described by a static wave function surrounding the nucleus as an "s orbital". Analytical solutions are available for a few simple cases, including a particle in a box and a quantum harmonic oscillator.

Example C.1. *Particle in a Box*
Consider a particle of mass m in a one-dimensional box that constrains the particle within the x-direction between $x = 0$ and $x = L$, with a zero potential energy inside the box and an infinite potential energy outside

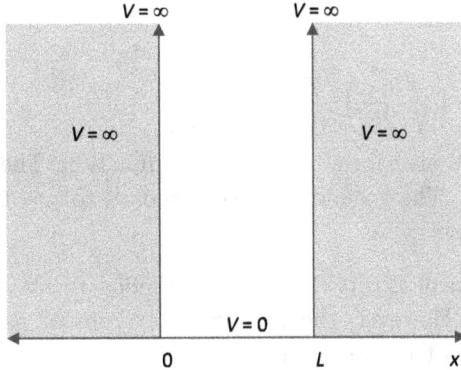

Figure C.1.   Particle in a one-dimensional potential energy box or infinite potential well.

of it (see Fig. C.1). The classical expression for the kinetic energy is $E = \frac{1}{2m}\hat{p}_x^2$, where $\hat{p}$ is the momentum of the particle. Thus, using the quantum momentum operator $\hat{p}_x = -i\hbar\frac{d}{dx}$, it therefore follows that the time-independent Schrödinger equation for this case is:

$$-\frac{\hbar^2}{2m}\frac{d^2\psi}{dx^2} = E\psi \tag{C.5}$$

The solution of this second-order ordinary differential equation with boundary conditions $\psi = 0$ at both $x = 0$ and $x = L$ is [Lewis *et al.*, 2022]:

$$\psi = A\sin(kx), \text{ where } k = \sqrt{2mE}/\hbar \quad \text{and} \quad k = \frac{n\pi}{L}, \ n = 1, 2.3... \tag{C.6}$$

As such, the energy levels are constrained to discrete values: $E_n = \frac{\hbar^2\pi^2 n^2}{2mL^2}$. A finite potential well is a generalization of this problem. A related problem is a rectangular well which gives rise to the concept of "quantum tunnelling" of particles through the potential well during radioactive decay.

Example C.2. *Harmonic oscillator*
For a harmonic oscillator, the Schrödinger equation is

$$-\frac{\hbar^2}{2m}\frac{d^2\psi}{dx^2} + \frac{1}{2}m\omega^2 x^2\psi = E\psi \tag{C.7}$$

where x is the displacement and ω is the angular frequency. This type of problem can describe vibrating atoms. It is the basis of a calculation perturbation method employed in quantum mechanics. The solution set for the eigenfunctions is

$$\psi_n(x) = \frac{1}{\sqrt{n!}} \left(\sqrt{\frac{m\omega}{2h}} \right)^n \left(x - \frac{\hbar}{m\omega} \frac{d}{dx} \right)^n \left(\frac{m\omega}{\pi\hbar} \right)^{1/4} e^{\frac{-m\omega x^2}{2\hbar}} \qquad (C.8)$$

which has a "Gaussian form" for the wave function. The eigenvalues are: $E_n = \left(n + \frac{1}{2} \right) \hbar\omega$. The ground state corresponds to $n = 0$, which is called the "zero-point energy".

The formalism of Eq. (C.1) can be extended to treat the presence of several particles. However, the number and types of particles are fixed. The Schrödinger equation cannot handle situations with changing particle number nor new types of particles appearing and disappearing.

A relativistic version of the Schrödinger equation, for example, in one dimension, is the "Klein–Gordon equation":

$$\left(\frac{1}{c^2} \frac{\partial^2}{\partial t^2} - \frac{\partial^2}{\partial x^2} + \frac{m^2 c^2}{\hbar^2} \right) \psi(x, t) = 0 \qquad (C.9)$$

This equation is "Lorentz-covariant" since the equation is true in every coordinate system. It is also a quantized version of the relativistic energy–momentum relation $E^2 = (pc)^2 + (m_0 c^2)^2$ in Appendix B.1. This equation pertains to spin-zero particles, i.e., it does not produce the fine structure for the hydrogen atom and allows negative energy states. Dirac subsequently published a relativistic quantum mechanics equation, relevant to spin-1/2 particles, using the square root of the relativistic energy–momentum relation:

$$i\hbar \frac{\partial \psi(x, t)}{\partial t} = \left[\beta m c^2 + c \sum_{n=1}^{3} \alpha_n p_n \right] \psi(x, t) \qquad (C.10)$$

In this expression, $\psi(x, t)$ is the wave function for an electron of mass m with space–time coordinates x, t. Here $\alpha_1, \alpha_2, \alpha_3$ and β are four 4×4 matrices for a four-component wave function ψ. The parameters p_1, p_2, p_3 are the components of the momentum, which are understood to be the "momentum operator" in the Schrödinger equation and c is the speed of light. The one-dimensional form of the free particle Dirac equation, for instance, is

$$i\hbar \frac{\partial \psi}{\partial t} = \left[m c^2 \boldsymbol{\sigma_0} - i\hbar c \boldsymbol{\sigma_1} \frac{\partial}{\partial x} \right] \psi \qquad (C.11)$$

where $\boldsymbol{\psi}$ is a two-component vector $\boldsymbol{\psi} = \begin{bmatrix} \psi_1 \\ \psi_2 \end{bmatrix}$ and the $\boldsymbol{\sigma}$'s are component matrices.

Equation (C.11), which is applicable to all fields associated to matter particles (i.e., with spin 1/2), can be written in the more compact form as shown in appendix D.3.1:

$$\boxed{\left(i\gamma^{\mu}\partial_{\mu} - m\right)\psi = 0}$$ (C.12)

Here ψ is the quantum field in four-dimensional space-time, which has four components and is known as a "*spinor*". The Dirac spinor encapsulates the intrinsic angular momentum (spin states) of fermions for fundamental particles like electrons and quarks in relativistic quantum mechanics. The parameter m is the mass of the particle, while ∂_{μ} denotes derivatives. The gamma matrices γ^{μ} in appendix D.3.1 are a set of 4×4 complex matrices that involve the anti-commutation relations:

$$\{\gamma^{\mu}, \gamma^{\nu}\} = 2\eta^{\mu\nu}I_4$$ (C.13)

in which $\eta^{\mu\nu}$ is the Minkowski metric element, where the indices run over 0,1,2 and 3 and I_4 is a 4×4 identity matrix.

C.1.1   *Uncertainty principle*

In classical mechanics, "Poisson brackets" are used to relate the canonical coordinates on phase space, which are used to describe a physical system in time. For instances, in classical mechanics, a typical example of canonical coordinates is q^i for the Cartesian coordinates of position and q_i for the components of momentum. Similarly, a corresponding "canonical commutation relation" can be formulated for quantum mechanics in an analogous fashion:

$$[\hat{x}, \hat{p}] = \hat{x}\hat{p} - \hat{p}\hat{x} = i\hbar$$ (C.14)

where $\hat{x}$ is a position operator and $\hat{p}$ is a momentum operator.

Computing "expectation values" for both $\hat{x}$ and $\hat{p}$, one can define an uncertainty for the observable as the standard deviation of both the position and momentum, respectively: $\sigma_x = \sqrt{\langle\hat{x}^2\rangle - \langle\hat{x}\rangle^2}$ and $\sigma_p = \sqrt{\langle\hat{p}^2\rangle - \langle\hat{p}\rangle^2}$. Consider the "Roberston uncertainty relation" as generally defined for any two "Hermitian operators" where

$$\sigma_A^2 \sigma_B^2 \geq \left(\frac{1}{2i}\langle[\hat{A}, \hat{B}]\rangle\right)^2$$ (C.15)

With $\hat{A} \equiv \hat{x}$ and $\hat{B} \equiv \hat{p}$, and using Eqs. (C.14) and (C.15), the uncertainty principle in quantum mechanics is formulated as

$$\boxed{\sigma_x \sigma_p \geq \left(\frac{1}{2i} \cdot i\hbar \right) = \left(\frac{\hbar}{2} \right)} \tag{C.16}$$

As an alternative derivation, one can consider the approach of Fourier that was derived about 100 years before the work of Heisenberg. This approach considered the superposition of waves with an uncertainty in position Δx and wave number Δk such that

$$\Delta x \Delta k \geq 1/2,$$

where the "wavenumber" $k = 2\pi/\lambda$ and λ is the "wavelength". Hence, based on the mathematical application of Fourier transforms that allows one to decompose a function in terms of its frequency components, this relation indicates that waves cannot have both a small size and number of frequencies. Moreover, in the case of a wave function in quantum mechanics, the momentum of the particle is given by $p = \hbar k$. Thus, multiplying both sides of the Fourier equation by $\hbar$ yields the uncertainty relation for the position and momentum of a particle in Eq. (C.16), i.e., $\Delta x \Delta p \geq \hbar/2$.

In quantum physics, Δp is no longer considered a range of spatial frequencies but rather an "uncertainty" in accordance with the Copenhagen interpretation that assumes a measurement will detect only one of the many waves of a pulse. Here the measurement disturbs the particle so that the wave function "collapses" and only one of its components is detected.

C.2 Quantum Field Theory

There are two approaches for carrying out calculations in quantum field theory:

(i) Canonical quantization,
(ii) Path integral or function quantization.

In quantum field theory, the word "field" specifically pertains to an operator field, which creates and destroys particle states. Four main components of quantum field theory include:

(i) Free fields and particles that are non-interacting with no forces involved. Solutions to the resultant equations are called "free-field solutions".

(ii) Interacting fields/particles. In practice, one uses the free-field solutions in item (i) and performs a perturbative calculation.

(iii) Renormalization is generally needed principally for the calculation of transition amplitudes, where some numerical factors are infinite, which can be made tractable with appropriate methods.

(iv) Application to experiment where determination of the interaction probabilities can be used to calculate scattering cross sections, decay probabilities/half lives among other experimental results.

Quantum mechanics only deals with a single particle such as an electron in, for instance, a potential well or in the case of a vibrating harmonic oscillator. There is no way generally to treat the transmutation of a particle such as the interaction of a particle with its anti-particle. In this reaction, the two particles annihilate one another to yield a neutral particle or the decay of an elementary particle. For instance, quantum field theory can describe a scattering process where an electron and positron produce a muon and anti-muon, i.e., $e^- + e^+ \rightarrow \mu^- + \mu^+$, as depicted in the "Feynman diagram" of Fig. C.2. As shown in this figure, at event x_2, the electron and positron annihilate one another creating a "transitory" photon represented by a wavy line. At the event x_1, this resulting photon is transmutated into a muon and anti-muon represented by straight lines with arrow heads pointing opposite their direction of travel through time. As shown, a standard convention in quantum field theory gives the numbering of events as going from $2 \rightarrow 1$. Unlike the real incoming and outgoing particles, the photon is not "real" but rather it is a "virtual particle" since it is undetectable and only mediates the interaction between the real particles.

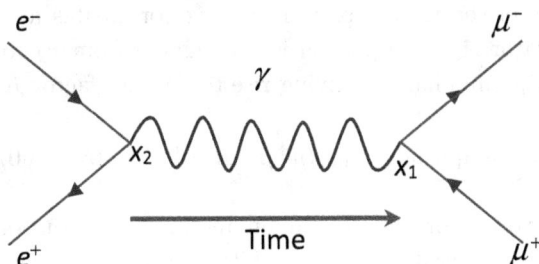

Figure C.2.   Scattering of an electron and positron that produces a muon and anti-muon.

In the general concept of quantum field theory as described by [Klauber, 2013], one has creation and annihilation operators acting on states (as represented by "kets"), yielding a "transition amplitude". The transition amplitude reflects the interaction process as shown, for example, in Fig. C.2. The transition amplitude for the interaction process in Fig. C.2 can be fundamentally described as:

Transition amplitude

$$= {}_{\text{final}}\langle\mu^+\mu^-|\underbrace{K}_{\substack{\text{const.}\\\text{theory}}}(\bar{\psi}_c^{\mu^-}\underbrace{\mathcal{A}_d\bar{\psi}_c^{\mu^+})(\bar{\psi}_d^{e^+}\mathcal{A}_c}_{\substack{\text{virtual photon}\\\text{which propagates}\\\text{from }x_2\text{ to }x_1}}\bar{\psi}_d^{e^-})|e^+e^-\rangle_{\text{initial}} \qquad (C.17)$$

Here the ket $|e^+e^-\rangle$ represents the incoming e^- and e^+, the bra $\langle\mu^+\mu^-|$ represents the outgoing muon and anti-muon, and the constant K is determined by theory. The operator $\bar{\psi}_d^{e^-}$ destroys the electron in the ket, the operator $\bar{\psi}_d^{e^+}$ destroys the positron, while the operator $\bar{\psi}_c^{\mu^+}$ creates an anti-muon in the ket and $\bar{\psi}_c^{\mu^-}$ creates a muon. The operator $\mathcal{A}_c$ creates a virtual photon while the operator $\mathcal{A}_d$ destroys the virtual photon. The under bracket shows the propagation process for the virtual (particle) photon from one event to another. It is called a "contraction", represented by a mathematical function known as a "Feynman propagator".

In this process, as an intermediate step, the incoming particles in the ket are destroyed by the destruction operators so that

$$\text{Transition amplitude} = {}_{\text{final}}\langle\mu^+\mu^-|K(\bar{\psi}_c^{\mu^-}\mathcal{A}_d\bar{\psi}_c^{\mu^+})(\mathcal{A}_c)K_2|0\rangle \qquad (C.18)$$

The destruction operators have acted on the original ket to leave a "vacuum state" $|0\rangle$ with no particles in it along with a simple numerical factor K_2. In the next step, the virtual photon propagator creates a virtual photon with the operator $\mathcal{A}_c$. The photon is propagated from x_2 to x_1 and then the operator $\mathcal{A}_d$ annihilates it giving rise to another factor K_γ:

$$\text{Transition amplitude} = {}_{\text{final}}\langle\mu^+\mu^-|K(\bar{\psi}_c^{\mu^-}\bar{\psi}_c^{\mu^+})K_\gamma K_2|0\rangle \qquad (C.19)$$

The remaining operators create a muon and anti-muon out of the vacuum, yielding a newly created ket $|\mu^+\mu^-\rangle$ with the numerical factor K_1. Since the remaining ket and bra represent the same multi-particle state, their inner product (bracket) is not zero. The combined quantity of factors can be moved outside of the bracket:

$$\text{Transition amplitude} = {}_{\text{final}}\langle\mu^+\mu^-|KK_1K_\gamma K_2|\mu^+\mu^-\rangle_{\text{final}}$$

$$= {}_{\text{final}}\langle\mu^+\mu^-|S|\mu^+\mu^-\rangle_{\text{final}}$$

$$= S_{\text{final}}\underbrace{\langle\mu^+\mu^-||\mu^+\mu^-\rangle_{\text{final}}}_{=1}$$

$$= S \qquad\qquad (C.20)$$

Here the quantity $S = KK_1K_\gamma K_2$, which depends on the particle momenta, spins, and masses as well as the strength of the electromagnetic interaction. The final probability of the interaction is equal to $S^\dagger S = |S|^2$. The scattering cross section for the interaction is calculated from the interaction probability.

C.2.1   *Quantization*

In quantum mechanics, solutions of the relevant wave equations are states. In quantum field theory however, the solutions are instead operators that create and destroy states. A "first quantization" was initially used to convert from classical theory to quantum particle theory called "particle quantization". The "second quantization" provided a means to convert to a field theory.

For consistency with special relativity, the wave function in the Klein–Gordon and Dirac equations was considered for the single particle states. In contrast, in field theory, it is field operators that create and destroy new particles. As such, one must replace the commutation relation for the operators in quantum mechanics with quantities specific for fields such that:

$$[\hat{x},\hat{p}] \rightarrow [\hat{\varphi}(\boldsymbol{x},t),\hat{\pi}(\boldsymbol{y},t)] \qquad\qquad (C.21)$$

The "field operator" $\hat{\pi}(\boldsymbol{x},t)$ corresponds to the "conjugate momentum density" of the field $\hat{\varphi}(\boldsymbol{x},t)$. However, it is noted that there is a continuum in quantum field theory as compared to discrete values in quantum mechanics. As such, in place of the discrete commutation relation in Eq. (C.14), one has an analogous commutation relation of the form:

$$[\varphi_r(\boldsymbol{x},t),\pi_s(\boldsymbol{y},t)] = i\hbar\delta_{rs}\delta(\boldsymbol{x}-\boldsymbol{y}) \qquad\qquad (C.22)$$

where δ is the Dirac delta function and $\boldsymbol{x}$ and $\boldsymbol{y}$ represent different three-dimensional position vectors. This representation respects the causality of special relativity so that if two fields are spatially separated, they do not affect one another. The operator π_s is the specific conjugate momentum density of the field φ_s. Different values for r and s mean different fields. Thus, in Eq. (C.22), the field dynamical variables become non-commuting operators. Space–time is therefore filled with fields $\varphi(x, t)$ that act as operators.

Although time t in quantum mechanics is just a parameter, the position, on the other hand, is an operator $\hat{x}$. However, in quantum field theory, one demotes position to a parameter x on equal footing with time. Thus, in summary, in quantum field theory, the fields $\hat{\varphi}$ and $\hat{\pi}$ are operators that are parametrized using numbers as points in space–time (x, t). The momentum continues to play the role of an operator.

As in classical mechanics and general relativity (see, for example, Appendices A.1.1 and B.2), the concept of the Lagrangian L, which represents the difference between the kinetic energy and potential energy in a system, i.e., $L = T - V$, is an important concept that arises in quantum field theory. This approach is used because important symmetries (such as rotations) in quantum field theory leave the Lagrangian invariant. Moreover, the classical (shortest) path that is taken by a particle, in accordance with the Lagrange–Euler equations in Eqs. (A.7) and (B.9), is one in which the action $S = \int L dt$ is minimized such that $\delta S = 0$.

A real scalar field $\phi(x, t)$ is the simplest type of classical field, in which there exists a real number at every point in space for the position vector x that can change as a function of time t. Consider a Lagrangian of this field of the form:

$$L = \int d^3x \mathcal{L} = \int d^3x \left[\frac{1}{2}\dot{\phi}^2 - \frac{1}{2}\left(\nabla\phi\right)^2 - \frac{1}{2}m^2\phi^2 \right] \qquad \text{(C.23)}$$

Here $\mathcal{L}$ is the Lagrangian density, $\dot{\phi}$ is the time-derivative of the field, ∇ is the gradient operator and m is the mass of the field. The Euler–Lagrange equation in Eq. (A.7) can be used to obtain the equation of motion for the field:

$$\frac{\partial}{\partial t}\left[\frac{\partial \mathcal{L}}{\partial(\partial\phi/\partial t)}\right] + \sum_{i=1}^{3}\frac{\partial}{\partial x^i}\left[\frac{\partial \mathcal{L}}{\partial(\partial\phi/\partial x^i)}\right] - \frac{\partial \mathcal{L}}{\partial \phi} = 0 \qquad \text{(C.24)}$$

Hence, the equations of motion for the field is simply the Klein–Gordon equation in Eq. (C.9):

$$\left(\frac{\partial^2}{\partial t^2} - \nabla^2 + m^2\right)\phi = 0 \qquad \text{(C.25)}$$

which is a wave equation. For discrete variables, a solution of the wave equation can be obtained through a Fourier series analysis. However, for the field in the continuum, one can instead use an analogue Fourier transform to sum over the normal modes for the oscillating system such that:

$$\phi(\mathbf{x}, t) = \int \frac{d^3p}{(2\pi)^3} \frac{1}{\sqrt{2\omega_\mathbf{p}}} \left(a_\mathbf{p} e^{-i\omega_\mathbf{p} t + i \mathbf{p} \cdot \mathbf{x}} + a_\mathbf{p}^* e^{i\omega_\mathbf{p}^* t - i \mathbf{p} \cdot \mathbf{x}} \right) \qquad (C.26)$$

Here $a_\mathbf{p}$ is a complex number (suitably normalized), the superscript "*" indicates the "complex conjugate" of $a_\mathbf{p}$, and $\omega_\mathbf{p}$ is the frequency of the normal mode defined as

$$\omega_\mathbf{p} = \sqrt{|\mathbf{p}|^2 + m^2} \qquad (C.27)$$

As such, each normal mode corresponding to a single $\mathbf{p}$ can be viewed as a classical harmonic oscillator with frequency $\omega_\mathbf{p}$. The displacement x of a classical harmonic oscillator from its equilibrium position is given by

$$x(t) = \frac{1}{\sqrt{\omega}} a e^{-i\omega t} + \frac{1}{\sqrt{\omega}} a^* e^{i\omega t} \qquad (C.28)$$

where a is a complex number (suitably normalized) and ω is the frequency of the oscillator. Similarly, for a quantum harmonic oscillator, $x(t)$ is promoted as a linear operator $\hat{x}(t)$:

$$\hat{x}(t) = \frac{1}{\sqrt{\omega}} \hat{a} e^{-i\omega t} + \frac{1}{\sqrt{\omega}} \hat{a}^\dagger e^{i\omega t} \qquad (C.29)$$

Here the complex numbers a and a^* are replaced by the the annihilation and creation operators $\hat{a}$ and $\hat{a}^\dagger$. The superscript symbol $\dagger$ indicates an operator that is a "Hermitian conjugate". It has the commutation relation $[\hat{a}, \hat{a}^\dagger] = 1$. This simple harmonic oscillator has the Hamilton

$$\hat{H} = \hbar\omega \hat{a}^\dagger \hat{a} + \frac{1}{2}\hbar\omega \qquad (C.30)$$

The vacuum state denoted as $|0\rangle$, which is the lowest energy state defined as $\hat{a}|0\rangle = 0$, has an energy of $\frac{1}{2}\hbar\omega$. Any energy eigenstate of a single harmonic oscillator can be obtained from $|0\rangle$ by successively applying the operator $\hat{a}^\dagger$, where the creation operator increases the energy of the oscillator by $\hbar\omega$. Hence, $\hat{a}^\dagger|0\rangle$ is an eigenstate of energy $3\hbar\omega/2$.

An analogous procedure can generally be applied to the real scalar field in Eq. (C.26). It can be promoted to a quantum field operator $\hat{\phi}$, with annihilation and creation operators and angular frequency pertaining to a particular $\mathbf{p}$:

$$\hat{\phi}(\mathbf{x}, t) = \int \frac{d^3 p}{(2\pi)^3} \frac{1}{\sqrt{2\omega_\mathbf{p}}} \left(\hat{a}_\mathbf{p} e^{-i\omega_\mathbf{p} t + i\,\mathbf{p}\cdot\mathbf{x}} + \hat{a}_\mathbf{p}^\dagger e^{i\omega_\mathbf{p}^* t - i\,\mathbf{p}\cdot\mathbf{x}} \right) \qquad \text{(C.31)}$$

The commutation relations are:

$$[\hat{a}_\mathbf{p}, \hat{a}_\mathbf{q}^\dagger] = (2\pi)^3 \delta(\mathbf{p} - \mathbf{q}), \ [\hat{a}_\mathbf{p}, \hat{a}_\mathbf{q}] = [\hat{a}_\mathbf{p}^\dagger, \hat{a}_\mathbf{q}^\dagger] = 0 \qquad \text{(C.32)}$$

The vacuum state is similarly defined as: $\hat{a}_\mathbf{p}|0\rangle = 0$ for all $\mathbf{p}$. The state space of a single quantum harmonic oscillator contains all the discrete energy levels of one oscillating particle. However, a quantum field contains discrete energy levels of an arbitrary number of particles (which are not fixed), which is known as a "Fock space". The process of quantizing an arbitrary number of particles, as opposed to just a single particle, is referred to as "second quantization".

C.2.2 *Feynman's path integrals*

The path integral formulation for quantum field theory provides a means for direct computation of the scattering amplitude of certain interaction processes in place of the use of operators and state spaces.

Feynman reasoned that a particle/wave such as an electron travelling a path between two events could actually be travelling along all possible paths (that are infinite in number) between events. A path integral is a way to calculate the amplitude of a system that starts off in some state $|\phi_I\rangle$ at time $t = 0$ and ends up in a final state $|\phi_F\rangle$ at time T by adding up amplitudes for the system to pass through all possible paths. One can split up the total time into N small intervals. The overall amplitude is the product of the amplitude of evolution within each interval integrated over all intermediate states. If the system dynamics are described by the Hamiltonian, then the amplitude is

$$\langle\phi_F|e^{-iHT}|\phi_I\rangle = \int d\phi_1 \int d\phi_2 ... \int d\phi_{N-1} \langle\phi_F|e^{-iHT}|\phi_{N-1}\rangle ...$$

$$\times \langle\phi_2|e^{-iHT}|\phi_1\rangle\langle\phi_1|e^{-iHT}|\phi_I\rangle$$

$$= \prod_{j=1}^{N-1} \int d\phi_j \langle\phi_F|e^{-iHT}|\phi_{N-1}\rangle ...$$

$$\times \langle\phi_2|e^{-iHT}|\phi_1\rangle\langle\phi_1|e^{-iHT}|\phi_I\rangle \qquad \text{(C.33)}$$

Taking the limit $N \to \infty$, the above product of integrals becomes (see [Klauber, 2013]):

$$\langle \phi_F | e^{-iHT} | \phi_I \rangle = \int_{\phi_I}^{\phi_F} D\phi(t) \exp \left\{ i \int_0^T L\, dt \right\} \tag{C.34}$$

where

$$D\phi = \lim_{N \to \infty} \prod_{j=1}^{N-1} \int d\phi_j C \tag{C.35}$$

Here C is a constant included in the definition of D. L is the Lagrangian, involving ϕ and its derivatives with respect to the spatial and time coordinates, where the Lagrangian is obtained from the Hamiltonian H through a Legendre transformation. The initial and final conditions of the path integral are respectively given by $\phi(0) = \phi_I$ and $\phi(T) = \phi_F$. Thus, in this formulation, the overall amplitude is obtained as the sum over the amplitude of every possible path between the initial and final states. The amplitude of a path is evaluated from the exponential of the action.

Example C.3. *Propagator for the Klein-Gordon Equation*
The path integral in Eq. (C.34) can be written as [Zee, 2010]:

$$Z = \int D\phi\, e^{i/\hbar \int d^4 x L(\phi)} \tag{C.36}$$

The functional integral in Eq. (C.36) is difficult to do except for Gaussian theory (i.e., a "free scalar field"), where $L(\phi) = \frac{1}{2}[(\partial \phi)^2 - m^2 \phi^2]$. Here the equation of motion is the Klein-Gordon equation in Eq. (C.25):

$$(\Box + m^2)\phi = 0 \tag{C.37}$$

where the d'Alembertian operator $\Box = \partial^\mu \partial_\mu$. The Feynman "propagator" for a free scalar field can be derived using a Green's function method and a Fourier transform technique. The equation for the Green's function $G_F(x - y)$ in the context of the inhomogeneous Klein-Gordon equation is ([Zee, 2010]):

$$(\Box + m^2)G_F(x - y) = -\delta^{(4)}(x - y) \tag{C.38}$$

where $\delta^{(4)}$ is four copies of the Dirac delta function. The Fourier transform is given by $\tilde{f}(k) \equiv \mathcal{F}\{f(x)\} = \int d^4 x\, e^{ik \cdot x} f(x) \Rightarrow \mathcal{F}\{\partial_\mu G(x)\} = \int d^4 x\, e^{ik \cdot x} \partial_\mu G(x)$. Using integration by parts and assuming that the boundary terms vanish:

$$\mathcal{F}\{\partial_\mu G(x)\} = -\int d^4x \; G(x) \; \partial_\mu \left[e^{ik\cdot x}\right]$$

$$= -\int d^4x \; G(x) \; ik_\mu e^{ik\cdot x} = -ik_\mu \tilde{G}(k)$$

Applying this method again for the second derivative yields:

$$\mathcal{F}\{\partial_\mu \partial^\mu G(x)\} = (-ik_\mu)(-ik^\mu)\tilde{G}(k) = -k^2 \tilde{G}(k)$$

Thus, for the property of Fourier transforms where differentiation in position space corresponds to multiplication by $(-ik_\mu)$ in momentum space, the Fourier transform to $\Box G(x-y)$ becomes $\Box G_F(x-y) \rightarrow -k^2 \tilde{G}(k)$. For the right hand side of Eq. (C.38), the Fourier transform of the $\delta^{(4)}$ is:

$$\int d^4x e^{ik\cdot x} \delta^{(4)}(x-y) = e^{ik\cdot y}$$

Here the delta function picks out the integrand at y. Combining, the transformed left and right hand sides and assuming that (y $=$ 0) since the origin does not affect the form of the Green's function in momentum space:

$$(-k^2 + m^2)\tilde{G}_F(k) = -1 \text{ or } (k^2 - m^2)\tilde{G}_F(k) = 1$$

Solving for $\tilde{G}_F(k)$ yields:

$$\tilde{G}_F(k) = \frac{1}{(k^2 - m^2)} \tag{C.39}$$

To ensure the correct casual behaviour, a small imaginary part $(i\epsilon)$ is inserted into the denominator:

$$\tilde{G}_F(k) = \frac{1}{(k^2 - m^2 + i\epsilon)} \tag{C.40}$$

Taking the inverse Fourier transform of Eq. (C.40), such that $f(x) = \int \frac{d^4k}{(2\pi)^4} e^{-ik\cdot x} \tilde{f}(k)$, yields the solution for the Feynman free propagator:

$$\boxed{D(x-y) \equiv G_F(x-y) = \int \frac{d^4k}{(2\pi)^4} \frac{e^{-ik\cdot (x-y)}}{k^2 - m^2 + i\epsilon}} \tag{C.41}$$

Using contour integration and the residue theorem [Lewis *et al.*, 2022], the integral in Eq. (C.41) can be evaluated. By splitting the integral up into spatial and temporal parts:

$$D(x-y) = \int \frac{d^3k}{(2\pi)^3} \int \frac{dk^0}{(2\pi)} \frac{e^{-ik^0 t} e^{ik\cdot (x-y)}}{(k^0)^2 - w_k^2 + i\epsilon} \tag{C.42}$$

where ω_k is defined as $\omega_k = \sqrt{\mathbf{k}^2 + m^2}$. One can use x^0 and t interchangeably in the nomenclature. The integrand has two poles in the complex-k^0 plane. To evaluate the integral with the residue theorem, the poles are shifted to $+\omega_k - i\epsilon$ and $-\omega_k + i\epsilon$. Here ϵ is a small number later taken in the limit to $\epsilon \to 0$. One pole is in the upper-half plane and the other in the lower-half plane (see Fig. C.3). The integral for a given pole *surrounded by a contour* is evaluated with a closed contour Γ along the real k^0 axis from $-\infty$ to $+\infty$ and an infinite semicircle in the $\text{Im}(k^0)$ half of the complex plane. In the upper-plane, the contour integral occurs in a counter-clockwise fashion around the pole, while in the lower-plane, it is in a counter-clockwise direction. Thus, integrating the temporal integral over k^0 with fixed k yields:

$$I_k(t, \omega_k) = \oint_\Gamma \frac{dk^0}{(2\pi)} \frac{e^{-ik^0 t}}{(k^0 - w_k + i\epsilon)(k^0 + w_k - i\epsilon)}$$

Outside of the real axis, the exponential term $e^{-ik^0 t}$ with positive t is exponentially damped for k^0 in the lower plane. Consequently, the residue theorem for the pole in the lower half-plane gives: $I(t, \omega_k) = -2\pi i \times$ Residue at $k^0 = +\omega_k - i\epsilon$ where the minus sign indicates a clockwise

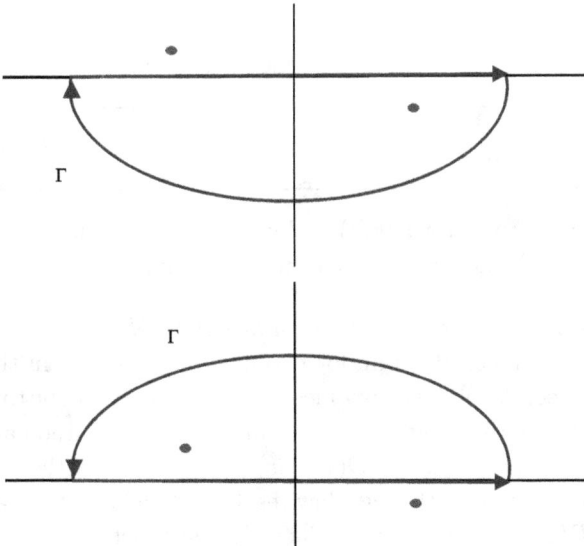

Figure C.3. Contour for integration of the Feynman propagator for a free field.

direction of the contour. Thus,

$$I_k(t, \omega_k) = -2\pi i \times \left(\frac{e^{-ik^0 t}}{(2\pi)(k^0 + w_k - i\epsilon)} \right)_{k^0 = +w_k - i\epsilon}$$

$$= -i \frac{e^{-i(w_k - i\epsilon)t}}{2(w_k - i\epsilon)} = -i \frac{e^{-i(w_k)t}}{2w_k}$$

where the last term is in the limit that ϵ approaches zero. Thus, the solution for $t > 0$ is:

$$D(x, t) = -i \int \frac{d^3 k}{(2\pi)^3} \frac{1}{2w_k} \exp i(\mathbf{x} \cdot \mathbf{k} - w_k t) \qquad \text{(C.43)}$$

Similarly, for $t < 0$, integrating over the positive half of the complex plane for the pole at $k^0 = -w_k + i\epsilon$ in a counter-clockwise direction with a plus sign yields:

$$I_k(t, \omega_k) = -i \frac{e^{+i(w_k)t}}{2w_k}$$

so that:

$$D(x, -t) = -i \int \frac{d^3 k}{(2\pi)^3} \frac{1}{2w_k} \exp i(\mathbf{x} \cdot \mathbf{k} + w_k t) \qquad \text{(C.44)}$$

Using the Heaviside step function θ, the combined integration results are:

$$\boxed{D(x, t) = -i \int \frac{d^3 k}{(2\pi)^3} \frac{1}{2w_k} \left[e^{i(\mathbf{x} \cdot \mathbf{k} - w_k t)} \theta(x^0) + e^{i(\mathbf{x} \cdot \mathbf{k} + w_k t)} \theta(-x^0) \right]} \qquad \text{(C.45)}$$

Physically, the propagator $D(x, t)$ evaluates the amplitude for a disturbance of the scalar field that propagates from the origin.

Example C.4. *Feynman Rules for a Scalar Field*
Consider the Feynman diagram for two mesons scattering in the left pane of Fig. C.4 [Zee, 2010]. The two mesons start from their birthplace at x_1 and x_2 and travel to a point in space-time w. The amplitude at this point for the Feynman propagator is $D(x_1 - w)D(x_2 - w)$. They then scatter with amplitude $-i\lambda$. After scattering, they reach their end points x_3 and x_4 with amplitude $D(x_3 - w)D(x_4 - w)$. There is a symmetry in the propagator where $D(x) = D(-x)$. The interaction point w could be anywhere in space-time. Feynman proposed a set of rules for evaluating the amplitude of the

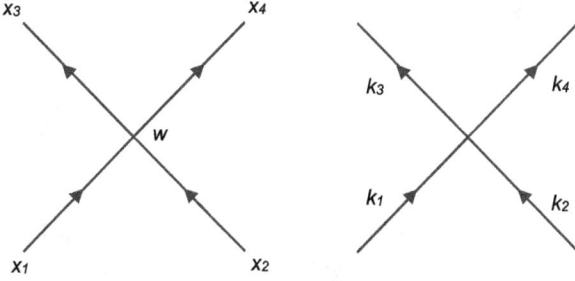

Figure C.4. Feynman diagrams for the scattering of two mesons (left pane) and momentum space (right pane).

event. For example, for the diagram in Fig. C.4, a factor of $-i\lambda$ is associated with the scattering event, and the factor $D(x_1 - w)$ for the propagation from x_1 to w and so on. In the momentum space of Fig. C.4, a meson with momentum k_1 and a meson with momentum k_2, collide and scatter. These particles emerge with momentum k_3 and k_4. The space-time propagator from Eq. (C.41) in Example C.3 can be written as:

$$D(x_a - w) = \int \frac{d^4 k_a}{(2\pi)^4} \frac{e^{\pm i k_a (x_a - w)}}{k_a^2 - m^2 + i\epsilon} \tag{C.46}$$

With dummy integration, there is a freedom of choosing either a plus or minus sign in the exponential. Integrating over w ([Zee, 2010]):

$$\int d^4 w \, e^{-i(k_1 + k_2 - k_3 - k_4)w} = (2\pi)^4 \delta^{(4)}(k_1 + k_2 - k_3 - k_4) \tag{C.47}$$

The signs indicate incoming and outgoing momentum, where for a conservation of momentum: $k_1 + k_2 = k_3 + k_4$. The Feynman diagrams are simply pictures of what happened.

Thus, the Feynman rules for scalar field theory are:

1. Draw a Feynman diagram of the process.
2. Label each line with momentum k and associate with it the propagator $i/(k^2 - m^2 + i\epsilon)$.
3. Associate each interaction vertex with the coupling $-i\lambda$, and $(2\pi)^4 \delta^{(4)} \sum_i k_i - \sum_j k_j$ thereby forcing a conservation of momentum.
4. Momentum associated with internal lines are to be integrated over with the measure $d^4/(2\pi^4)$.
5. Some diagrams are to be multiplied by a symmetry factor such as $1/2$.

Applying the rules for the diagram in the right pane of Fig. C.4 yields the amplitude M:

$$M = -i\lambda(2\pi)^4 \delta^{(4)}(k_1 + k_2 - k_3 - k_4) \prod_{a=1}^{4} \left(\frac{i}{k_a^2 - m^2 + i\epsilon} \right)$$

The product factor in the last term is common to all diagrams, where two mesons scatter into two mesons. Thus, an additional Feynman rule 6 cancels the association of a propagator with *external lines* since the external particles satisfy $k_a^2 - m^2$. This rule is known as *amputating the external legs*. Also, since there is an overall factor for momentum conservation, the delta function is understood as Feynman rule 7. Therefore, applying all of these rules together, the above amplitude reduces to $M = -i\lambda$!

Consider now the case of two colliding mesons producing four mesons in the Feynman diagram of Fig. C.5. Amputating the external legs for the six external lines, keeping only the propagator for the *internal line* q (i.e., which, from a conservation of momentum, is $(k_4 + k_5 + k_6)$), putting in a factor $(i\lambda)$ for each vertex and ignoring the delta function, yields the amplitude:

$$M = (-i\lambda)^2 \frac{i}{(k_4 + k_5 + k_6)^2 - m^2 + i\epsilon}$$

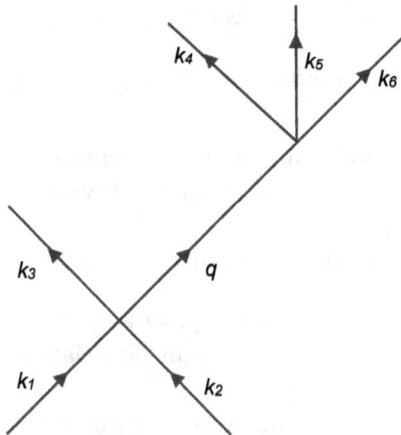

Figure C.5. Feynman diagrams for the scattering of two colliding mesons producing four mesons.

C.3   Loop Quantum Gravity

Loop quantum gravity provides a framework to merge quantum mechanics with Einstein's theory of general relativity as a theory of quantum gravity. A quantum theory of gravity is based on Einstein's geometric formulation rather than considering gravity as a force. The mathematical framework of this theory considers Ashtekar variables, which are a set of complex variables that describe the geometry of space–time at the Planck scale. The theory is formulated using non-perturbative techniques of loop quantum geometry, spin networks and knot theory. However, this theory is still evolving.

The mathematical foundations of loop quantum gravity for study of the quantum properties of the gravitational field and the geometry of space–time at the Planck scale include:

(i) *Ashtekar connection*: The Ashtekar connection is a complex variable defined as:

$$A_a^i = \Gamma_a^i + iK_a^i$$

where Γ_a^i is the Christoffel symbol for the space–time connection and K_a^i is the extrinsic curvature of a spatial surface. A connection encodes the parallel transport of vectors from one point in space–time to another. Hence, this connection specifically describes the geometry of space–time at the Planck scale. It enables a way to pull-back the space–time connection onto a three-dimensional surface. It therefore provides a means to define the quantum states of the gravitational field.

(ii) *Quantum Hamiltonian constraint*: This constraint describes the quantum evolution of the gravitational field, which governs the quantum dynamics of the Ashtekar connection. The Hamiltonian constraint relates the spatial metric of a quantum space–time to its matter content. It can be written in terms of "holonomies" that describe the parallel transport of the Ashtekar connection along a curve in space–time:

$$H_Q[N] = \int d^3x N(x)\widehat{H}_{Q(x)} = \int d^3x N(x)\left[\det(q)H_G + H_M\right]$$

Here $N(x)$ is a lapse function, $\widehat{H}_{Q(x)}$ is the quantum operator for the Hamiltonian constraint, and q is the spatial metric. The quantity H_G

is the gravitational part of the Hamiltonian:

$$H_G = -G^{ijkl} F_{ij}^a F_{kl}^b E_a^i E_b^j$$

where G^{ijkl} is the inverse of the Dewitt metric, F_{ij}^a is the curvature of the Ashtekar–Barbero connection, and E_a^i is the desensitized triad field. This latter field describes the spatial geometry of space–time at a quantum level modified so that it is less sensitive to quantum fluctuations. The parameter H_M is the matter part of the Hamiltonian given by the standard expression for matter fields in general relativity. This is a simplified version of the Hamiltonian constraint.

(iii) *Holonomy*: A holonomy is used to define the quantum states of the gravitational field and to calculate transition amplitudes between different states. The holonomy of the Ashtekar connection along a curve C is given by

$$h_C[A] = P \exp \left(\int_C A_a \, dx^a \right)$$

where P denotes the path ordering and the integral is taken along the curve C.

(iv) *Spin network*: A spin network is a two-dimensional surface that is embedded in four-dimensional space–time. This network is used to describe the quantum geometry of space–time. A spin network is a graph built from nodes and edges, each labelled by a representation of the gauge group SU(2) and an integer, respectively. It thereby represents quantum states of the gravitational field. The state of a spin network derived as a function on the nodes and edges can be used to calculate observables in loop quantum gravity.

(v) *Spin foam model*: A spin foam model provides a framework to describe the quantum geometry of space–time in terms of two-dimensional surfaces (i.e., spin networks) that are embedded in four-dimensional space–time. The spin foam model is used to compute transition amplitudes between different states of the gravitational field for calculation of physical quantities, such as the spectrum of geometric operators and the area and volume of space–time regions.

D   The Standard Model

The complete standard model is depicted in Fig. D.1. The names, masses, spins, charges, chiralities (i.e., left- and right-handedness) and interactions with the strong, weak and electromagnetic forces are summarized in this figure and explained in the following text.

The standard model is a "gauge quantum field theory" with an internal symmetry of the SU(3) × SU(2) × U(1) unitary product group. It describes fundamental particle physics of three generations of quarks and leptons, gauge bosons and the Higgs boson. This composite model embodies a fermion field (ψ) (which accounts for matter particles), electroweak boson fields (W_1, W_2, W_3) and B, a gluon field (G_a) and a "Higgs field" (φ) that are all operator valued in accordance with the requirements of quantum field theory. As depicted in the figure, "symmetry breaking" is an important phenomenon for the electroweak force in which the Higgs boson plays a crucial role. In the process of symmetry breaking, the Higgs field has a vacuum expectation value (VEV) (i.e., with a non-zero vacuum energy of 246 GeV). This value sets the scale for the mass of the observed particles. In addition, the chiral and gluon condensates in quantum chromodynamics theory provide mass to the quarks (i.e., hadrons). The hadrons are composite subatomic particles made up of two separate families of particles composed of baryons containing an odd number of quarks (such as protons and neutrons) and mesons that contain an even number of quarks (such as pions and kaons). As shown in the figure, and mathematically detailed in Appendix D.1, particles differ with the consequence of symmetry breaking from a high-energy symmetric state (upper top schematic of the figure) that occurs early in the creation of the universe to a low-energy broken symmetry phase (in the bottom schematic of the figure).

The left and right-handedness depicted in Fig. D.1 depends on the orientation of the spin. For instance, solving the Dirac equation in

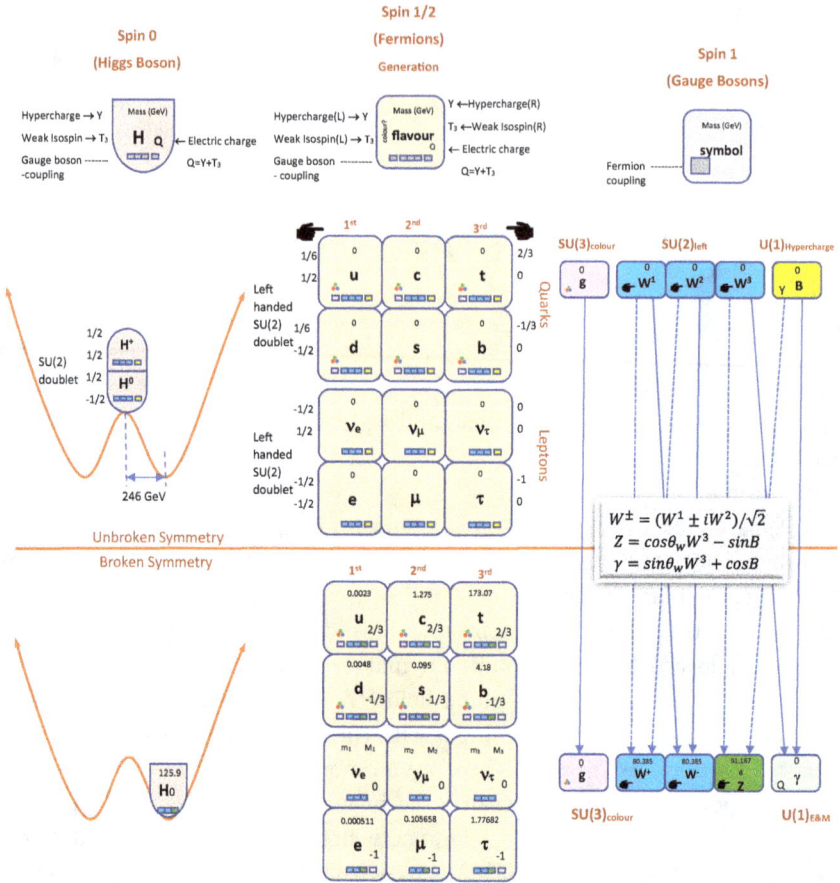

Figure D.1. Standard theoretical model of particle physics.

Eq. (C.12), as the speed of the particle increases, its spin becomes oriented along the direction of motion. Initially, all matter particles are massless in Fig. D.1 before symmetry breaking. Thus, for massless particles, which travel at the speed of light, their spin points either in the direction of motion (i.e., *right-handed* particle) or in the opposite direction (i.e., *left-handed* particle). The *helicity* quantifies this distinction, where for a particle of momentum $\mathbf{p}$ with a spin pointing in direction $\mathbf{s}$, the helicity is defined as: $h = (\mathbf{p} \cdot \mathbf{s})/|\mathbf{p}|$. Right-handed particles have a helicity of $+1/2$ and left-handed particles a value of $-1/2$.

Initially, at high energies in the electroweak theory, the universe has four components of the Higgs scalar field whose interactions are carried by four gauge bosons that are massless — each one similar to the photon forming a complex scalar Higgs field doublet (with neutral and charged components). These bosons possess a weak isospin SU(2) symmetry. The potential for this field is a "Mexican hat shape" with a non-zero VEV everywhere. A key feature of this field is that it takes less energy for the field to have a non-zero value than a zero value. This non-zero value in fact breaks the electroweak symmetry. In addition, there are four massless electroweak bosons shown in Fig. D.1. However at low energies, this symmetry spontaneously breaks down to the U(1) symmetry of electromagnetism when one of the Higgs fields acquires a vacuum expectation value. The three extra bosons become incorporated into the three weak bosons that acquire mass through the "Higgs mechanism" as detailed in Appendix D.2. These three bosons are the W^+, W^- and Z^0 particles observed with the weak interaction. The fourth electroweak gauge boson is the photon (γ). The photon does not couple to any of the Higgs field and remains massless. Thus, with spontaneous symmetry breaking, these fermions acquire a mass proportional to the Higgs field through a so-called "Yukawa interaction", which describes the coupling between the Higgs field and the massless quark and lepton fields.

Originally, Yukawa recognized that the potential energy between particles of mass m would vary with the separation distance r between them as:

$$V(r) \sim \frac{e^{-r/R}}{r} \tag{D.1}$$

Here the range R of the potential is inversely related to the mass: $R = \hbar/(mc)$ in accordance with the definition of the Compton wavelength $\lambda = \hbar c/E$ for a process of energy E. For instance, for the strong force with a range of $R \approx 2 \times 10^{15}$ m, a particle approximately 200 times heavier than the electron (or a mass of about 100 MeV) is required for the binding of neutrons and protons in the nucleus with the discovery of the pion in 1947 (see Section 1.1.1). The characteristic energy scale of QCD for the strong coupling scale is: $\Lambda_{QCD} \sim 200$ MeV. The three quarks (up, down and strange) in Fig. D.1 have a mass smaller than Λ_{QCD}, while the three other ones have masses much heavier.

The weak mixing angle or Weinberg angle, θ_W, in Fig. D.1, derived in the Weinberg-Salam theory of the electroweak interaction on spontaneous symmetry breaking, is the ratio of the coupling strengths between the weak and electromagnetic forces. The relationship between the masses of the W and Z bosons and the weak mixing angle can be visualized as a two-dimensional plane known as a "vector boson plane". On spontaneous symmetry breaking, the original W^0 and B^0 vector boson plane is rotated by this angle, thereby producing as a result the Z^0 particle and photon.

Symmetry breaking is not unique to particle physics but it is a common phenomenon also observed in condensed matter physics. In particle physics, the weak iospin symmetry of the electroweak interaction is broken by a Higgs mechanism. The "Goldstone bosons" that result from symmetry breaking are absorbed by massless W and Z bosons. These SU(2) and U(1) gauge bosons now become the massive W and Z bosons of the weak force through a Higgs mechanism. The remaining electrically neutral component manifests itself as a separate Higgs boson. This boson can also couple separately to the fermions of matter (through a Yukawa coupling) causing these particles to acquire mass as well.

The strong and weak nuclear forces are both carried by a field of spin 1. The Maxwell equations in Eq. (A.38) describe this field for electromagnetism. The strong and weak nuclear forces are described as a generalization of the Maxwell equations known as "*Yang-Mills equations*". In the Yang-Mills theory, the gauge group determines the symmetry properties of the theory that include the SU(3) group for quantum chromodynamics (QCD) and the SU(2) × U(1) group for the electroweak interaction. The gauge fields are also known as gauge bosons that are the carriers of the fundamental forces. These latter equations are in a similar mathematical form to the Maxwell equations (compare with Eq. (A.38)) and are given as:

$$D^\mu F^a_{\mu\nu} = J^a_\nu \tag{D.2}$$

The index a runs over the generators of the Lie algebra of the gauge group. Each component of Yang-Mills theory is a $N \times N$ matrix in space-time. The term $F^a_{\mu\nu}$ is the field strength tensor for a non-Albelian gauge field defined as: $F^a_{\mu\nu} = \partial_\mu A^a_\nu - \partial_\nu A^a_\mu + g f^{abc} A^b_\mu A^c_\nu$ where $A_\mu = A^a_\mu T^a$ and T^a are the generators of the SU(2) and SU(3) group. The parameter g is the gauge coupling constant and f^{abc} are the structure constants of the gauge group. In QCD, the gauge bosons are gluons while in electroweak theory they include the W and Z bosons and photons. These equations describe how the gauge fields (A^a_μ) evolve in the presence of an external source. The source term J^a_ν introduces an inhomogeneous term into the equation, analogous to the presence of a charge and current in Maxwell's equations.

Here D_μ is like a partial derivative with respect to space-time but it also includes some commutator properties:

$$D_\mu = \partial_\mu - igA_\mu \qquad (D.3)$$

where the covariant derivative acts on a field ϕ as:

$$(D_\mu\phi)^a = \partial_\mu\phi^a + gf^{abc}A_\mu^b\phi^c \qquad (D.4)$$

As mentioned, the SU(3) group describes the strong force, while SU(2) accounts for the weak force. However, U(1) does not pertain to the force of electromagnetism. Instead it describes an "electro-magnetism like" force called *"hypercharge"*. Thus, electromagnetism is not one of the three fundamental forces of the Standard Model. Rather it is a combination of the weak force and hypercharge, SU(2) × U(1), which collectively is referred to as the *electroweak* theory.

The hypercharge Y is a quantum number in the classification and interaction of particles. The hypercharge is related to the electric charge (Q) and the third component of isospin (T_3) by the Gell-Mann-Nishijima relation ($Q = T_3 + Y/2$). Moreover, particles are grouped into multiplets based on their isotopic spin (i.e., isospin) (I) where each multiplet contains particles with similar masses but different charges. For example, the proton and neutron form a doublet with $I = 1/2$, whereas the pions form a triplet with $I = 1$. The isospin projection (I_3 or T_3) is the third component of the isospin vector that determines the specific charge state within a multiplet. This projection is analogous to the m quantum number in quantum mechanics. It therefore explains the symmetries between hadrons and how they behave similarly despite having differing charges. For a multiplet with isospin I, I_3 has $(2I + 1)$ possible values, ranging from $-I$ to $+I$ in integer steps. Thus, for protons: ($I_3 = +1/2$ and $Q = 1$) while for neutrons: ($I_3 = -1/2$ and $Q = 0$).

The Lagrangian function accounts for the kinetic energy minus the potential energy of a field. It plays a key role in the standard model (see Appendix D.3), where it is applied in the Euler-Lagrange equations to evaluate the equations of motion of the field. This approach involves a "perturbative" calculation. In this approach, the Lagrangian is decomposed into a separate "free field" (subscript 0) and a "perturbative interaction component" (subscript I) such that: $\mathcal{L} = \mathcal{L}_0 + \mathcal{L}_I$. The free field term accounts for particles in isolation and the other component involves particle interactions. The free field for the fermions ψ obey the "Dirac equation" and the photon field A obeys the "wave equation", while the Higgs field

φ satisfies the "Klein-Gordon equation" (see Appendix C). The weak interaction fields for Z, $W^{\pm}$ satisfy the "Proca equation" for these massive spin-1 particles.

Mathematical formulations for spontaneous symmetry breaking (Appendix D.1) and the Higgs mechanism (Appendix D.2) are developed below to demonstrate how the masses of the fundamental particles in Fig. D.1 arise in the model. Lagrangian functions for the electromagnetic, weak and strong forces are given in Appendix D.3.

D.1   Spontaneous Symmetry Breaking

Recall that the Lagrangian is generally defined as $L = T - V$ where T is the kinetic energy and V is the potential energy. The Lagrangian density function for the Klein-Gordon relativistic wave equation for field quanta φ for particles of mass m is

$$\mathcal{L} = \frac{1}{2} \left(\partial_\mu \varphi \partial^\mu \varphi \right) - \frac{1}{2} m^2 \varphi^2 \tag{D.5}$$

This equation can be further abbreviated as

$$\mathcal{L} = \frac{1}{2} \left(\partial_\mu \varphi \right)^2 - \frac{1}{2} m^2 \varphi^2 \tag{D.6}$$

As such, the term $\frac{1}{2} \left(\partial_\mu \varphi \right)^2$ is the kinetic energy and the potential energy term is $\frac{1}{2} m^2 \varphi^2$. This Lagrangian can be generalized further using so-called "φ^4 theory". This approach becomes important in the idea of spontaneous symmetry breaking where

$$\mathcal{L} = \frac{1}{2} \left(\partial_\mu \varphi \right)^2 - \frac{1}{2} m^2 \varphi^2 - \frac{1}{4} \lambda \varphi^4 \tag{D.7}$$

and the parameter λ is the coupling strength. However, following [Schwartz, 2014], a system for instance can undergo symmetry breaking at a critical temperature T_C where $m^2(T) = \alpha (T - T_C)$ for some constant α. Then for $T > T_C$, the mass term $m^2 > 0$, and the Lagrangian describes an ordinary scalar field theory. However, for $T < T_C$, $m^2 < 0$. A negative mass-squared implies a tachyon particle with a momentum that is spacelike that communicates at a speed faster than light. For $T < T_C$, the field cannot be treated as a small excitation by expanding about $\varphi = 0$ in accordance with perturbative quantum field theory. Instead, one needs to consider excitations around the true vacuum. Thus, one can replace $m^2 \to -m^2$ so that m^2 is still positive and the Lagrangian becomes:

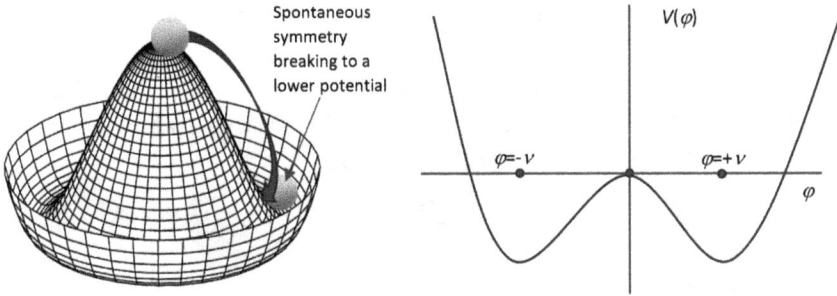

Figure D.2.   The potential for the Lagrangian for φ^4 theory.

$\mathcal{L} = \frac{1}{2}\left(\partial_\mu\varphi\right)^2 + \frac{1}{2}m^2\varphi^2 - \frac{1}{4}\lambda\varphi^4$. The potential is now minimized when φ has a constant non-zero value.

The potential is therefore recognized as $V(\varphi) = -\frac{1}{2}m^2\varphi^2 + \frac{1}{4}\lambda\varphi^4$. Taking the derivative of this function with respect to the variable φ and setting this result equal to zero gives the minimum of this potential function V. The result is shown graphically in Fig. D.2, where $\varphi = 0$ (i.e., the trivial ground state) and $\varphi = \pm\sqrt{\frac{m^2}{\lambda}} = \pm\nu$. In accordance with a "Mexican hat model" for the potential energy in the φ^4 theory, the case for which $\varphi = 0$ corresponds to the situation in which a particle is sitting on top of the hat. For the true minimum, one has $\varphi = +\nu$ or $\varphi = -\nu$. The symmetry can therefore be broken as the particle rolls down from the unstable point $\varphi = 0$ and into one of these lower minimum positions in the brim of the hat.

The minimum is located at $\varphi = \pm\nu$. However, one can rescale the field about the right-handed minimum ν as

$$\varphi(x) = \nu + \eta(x) \tag{D.8}$$

so that the minimum occurs at the position $\varphi = 0$ with field fluctuations described by $\eta(x)$. Substituting Eq. (D.8) into Eq. (D.7) gives the rescaled Lagrangian using $m^2 = \lambda\nu^2$ for the minimum of the potential and ignoring the constant term (i.e., $(1/4)\lambda\nu^4$) that does not contribute to the field equations:

$$\mathcal{L} = \frac{1}{2}\left(\partial_\mu\eta\right)^2 - \lambda\nu^2\eta^2 - \lambda\nu\eta^3 - \frac{1}{4}\lambda\eta^4 \tag{D.9}$$

On comparison of Eq. (D.9) with Eq. (D.6), it is seen that the first term in Eq. (D.9) is a kinetic energy term. The second term is a mass term, where $m = \sqrt{2\lambda}\nu$ as follows on comparison of the mass term in Eq. (D.9) with that in the Klein–Gordon Lagrangian of Eq. (D.6). The last two terms are self-interaction terms.

In the real physical situation, the spontaneous symmetry breaking will involve more than one particle. Thus, as an illustrative example, consider a complex field with two real fields φ_1 and φ_2:

$$\varphi = \frac{\varphi_1 + i\varphi_2}{\sqrt{2}} \tag{D.10}$$

with the Lagrangian:

$$\mathcal{L} = \left(\partial_\mu \varphi^\dagger \partial^\mu \varphi\right) - m^2 \varphi^\dagger \varphi + \lambda(\varphi^\dagger \varphi) \tag{D.11}$$

The complex conjugate of Eq. (D.10) is

$$\varphi^\dagger = \frac{\varphi_1 - i\varphi_2}{\sqrt{2}}$$

(so that $\varphi^\dagger \varphi = \left(\frac{\varphi_1 - i\varphi_2}{\sqrt{2}}\right)\left(\frac{\varphi_1 + i\varphi_2}{\sqrt{2}}\right) = \frac{1}{2}\left(\varphi_1^2 + \varphi_2^2\right)$). Substituting in φ and $\varphi^\dagger$ into Eq. (D.11) gives

$$\mathcal{L} = \frac{1}{2}\left(\partial_\mu \varphi_1\right)^2 + \frac{1}{2}\left(\partial_\mu \varphi_2\right)^2 - V \tag{D.12}$$

where the potential $V = \frac{1}{2}m^2\left(\varphi_1^2 + \varphi_2^2\right) - \frac{1}{4}\lambda\left(\varphi_1^4 + \varphi_2^4\right) - \frac{1}{2}\lambda\varphi_1^2\varphi_2^2$. This Lagrangian has rotational symmetry in φ_1 and φ_2. As such, the minimum of the potential V lies on a circle

$$\varphi_1^2 + \varphi_2^2 = \frac{m^2}{\lambda} \tag{D.13}$$

As shown previously, to break the U(1) rotational symmetry of a particle sitting in an unstable position at the top of the potential, one can chose the minimum at the lower potential position for φ_1 and φ_2:

$$\nu_1 = \frac{m}{\sqrt{\lambda}} \quad \text{and} \quad \nu_2 = 0 \tag{D.14}$$

The minima in Eq. (D.14) satisfies Eq. (D.13). Again, by redefining the fields with a coordinate shift so that they fluctuate around the minimum of Eq. (D.14):

$$\chi = \varphi_1 - \frac{m}{\sqrt{\lambda}} \quad \text{and} \quad \psi = \varphi_2 \tag{D.15}$$

Finally, solving for φ_1 and φ_2 in Eq. (D.15), and inserting these expressions into the Lagrangian of Eq. (D.12), one can obtain the *free* Lagrangian component which ignores particle interaction. Thus, one can ignore terms of the form φ^n where $n > 2$. For the free mass term, it is noted that $\frac{1}{2}m^2\left(\varphi_1^2 + \varphi_2^2\right) - \frac{1}{2}\lambda\varphi_1^2\varphi_2^2 = \frac{1}{2}m^2\chi^2$. Hence, the free part of the Lagrangian in Eq. (D.12) follows directly as

$$\mathcal{L}_{free} = \frac{1}{2}\left(\partial_\mu\chi\right)^2 + \frac{1}{2}\left(\partial_\mu\psi\right)^2 - \frac{1}{2}m^2\chi^2 \tag{D.16}$$

This derivation shows that with spontaneous symmetry breaking of the U(1) symmetry, a field χ with mass m $(= \nu_1\sqrt{\lambda})$ and a field ψ (which is *massless*) results. This example is for scalar fields for particles with a spin of 0. It also shows, importantly, how the massless "Goldstone boson" of spin 0 appears in the standard model as a consequence of symmetry breaking.

D.2   The Higgs Mechanism

The previous example involved spontaneous symmetry breaking for a U(1) symmetry in the case of a complex field with two real components in Eq. (D.10) under a *global* gauge invariance. This calculation can be extended further to the situation for a gauge field A_μ with a *local* gauge invariance under a U(1) transformation. The symmetry breaking of a massless gauge field A_μ into a vector field that contains mass explains the appearance of massive vector bosons $W^\pm$ and Z^0 in the electroweak theory. As a consequence of symmetry breaking, masses appear when the symmetry is broken. In this case, the electron, W and Z particles that mediate the weak force acquire mass. The Higgs field is actually a pair of complex-valued Higgs fields $\varphi(x) = \begin{pmatrix} \varphi_1 \\ \varphi_2 \end{pmatrix}$, where the two fields are mixed by the weak force. Since there are two Higgs fields, the potential $V(\varphi)$ takes on the 3-D shape shown by the Mexican hat model of Fig. D.2. In fact, the mechanism of spontaneous symmetry breaking with a gauge field and local U(1) invariance is the well known "Higgs mechanism" of particle physics.

The Lagrangian is invariant under a U(1) rotation $\varphi \rightarrow \varphi' = e^{-i\theta}\varphi$ where the angle θ is a scalar. This can be generalized further as $\varphi \rightarrow \varphi' = e^{-iq\theta(x)}\varphi$ where q is a number but $\theta(x)$ is a function of space–time thereby varying from point to point. For invariance of the Lagrangian, one requires the use of a "gauge field". By analogy with electromagnetics under a U(1) gauge transformation with a vector potential A_μ, one can use a transformation of the form: $A_\mu \rightarrow A'_\mu = A(x)_\mu + \partial_\mu\theta(x)$. In this case, the ordinary derivatives in the kinetic term of the Lagrangian $\partial_\mu\varphi^\dagger\partial^\mu\varphi$ must be replaced with a covariant derivative

$$D_\mu = \partial_\mu + iqA_\mu$$

The Lagrangian therefore becomes:

$$\mathcal{L} = D_\mu \varphi^\dagger D^\mu \varphi - V(\varphi^\dagger \varphi) - \frac{1}{4} F_{\mu\nu} F^{\mu\nu} \qquad (D.17)$$

Thus, the Lagrangian includes a complex scalar field and a massless gauge field A_μ. The term $F_{\mu\nu} F^{\mu\nu}$ corresponds to kinetic energy where, in analogy to electrodynamics (see Appendix A.2),

$$F_{\mu\nu} = \partial_\mu A_\nu - \partial_\nu A_\mu$$

Since there are no quadratic terms in the Lagrangian (i.e., $A_\mu A^\nu$) that would indicate mass as seen previously, the gauge field itself is initially massless. The potential energy term is given by

$$V(\varphi^\dagger \varphi) = \frac{m^2}{2\nu^2} (\varphi^\dagger \varphi - \nu^2)^2 \qquad (D.18)$$

Here ν is a real number for the minimum of the potential energy with an unbroken symmetry such that $|\varphi|^2 = \nu$. However, on symmetry breaking, as with the previous example, one seeks a minimum potential energy when both the gauge field A_μ and potential V vanish. Interestingly, a local gauge transformation provides the same minimum and symmetry as follows with the earlier transformation law: $\varphi'^\dagger \varphi' = (\varphi^\dagger e^{iq\theta(x)})(e^{-iq\theta(x)}\varphi) = \varphi^\dagger \varphi = |\varphi|^2 = \nu$. Again, one is interested in a gauge transformation that gives a field in terms of fluctuations around ν. Thus, for a vacuum ν perturbed by a real field, $h(x)$ (i.e., the so-called Higgs field):

$$\varphi \to \varphi' = \nu + \frac{h(x)}{\sqrt{2}} \qquad (D.19)$$

Since the Higgs field is real, it follows from Eq. (D.19) that the field itself is real where $\varphi^\dagger = \varphi$. Substituting φ' in Eq. (D.19) into Eq. (D.18) gives

$$V = \frac{m^2}{2\nu^2} \left[\sqrt{2}\nu h(x) + \frac{h^2(x)}{2} \right]^2$$

Using Eq. (D.19) along with the covariant derivative expression gives

$$D_\mu \varphi' D^\mu \varphi' = \left[(\partial_\mu - iqA') \left(\nu + \frac{h}{\sqrt{2}} \right) \right] \cdot \left[(\partial_\mu + iqA') \left(\nu + \frac{h}{\sqrt{2}} \right) \right]$$

Inserting the last two expressions into Eq. (D.17) gives the final result for the complete Lagrangian expression (dropping the primes on the vector potential for clarity):

$$\mathcal{L} = \underbrace{\frac{1}{2}\partial_\mu h \partial^\mu h - m^2 h^2}_{L^h_{free}} + \underbrace{q^2 \nu^2 A_\mu A^\mu - \frac{1}{4}F_{\mu\nu}F^{\mu\nu}}_{L^B_{free}}$$

$$\underbrace{-\frac{m^2 h^2}{2\nu^2}\left(\sqrt{2}\nu h + \frac{h^2}{4}\right)}_{L^h_{int}} + \underbrace{q^2 A_\mu A^\mu \left(\sqrt{2}\nu h + \frac{1}{2}h^2\right)}_{L^{coup}_{int}} \qquad \text{(D.20)}$$

The free part of the Lagrangian involving the Higgs field in Eq. (D.20), L^h_{free}, is simply a Lagrangian for a Klein–Gordon type equation with a scalar field $h(x)$ of mass $\sqrt{2}m$. Hence, the scalar Higgs field is a spin-0 boson of mass $\sqrt{2}m$. The free Lagrangian for the gauge field, L^B_{free}, is unique. There is a kinetic energy term as in the original Lagrangian. However, before symmetry breaking, the gauge field was massless. On choosing a real field with a perturbation about the unbroken vacuum ν, a mass term $q^2\nu^2 A_\mu A^\mu$ is now remarkably picked up! Comparing this latter term to one that would appear in a Klein–Gordon type Lagrangian (i.e., $\frac{1}{2}M\varphi^2$), the process of symmetry breaking has given rise to a vector boson with a mass of $M = \sqrt{2}q\nu$. The remaining terms in the Lagrangian are interaction terms. These include a self-interaction term for the Higgs field (i.e., L^h_{int}), and an interaction term accounting for the coupling between the Higgs field h and the gauge field A_μ (i.e., L^{coup}_{int}).

In summary, a spontaneous symmetry breaking process has led to the appearance of massive particles in the Lagrangian. In this process, a Lagrangian is considered for a vacuum state, after which, with spontaneous symmetry breaking, a new vacuum state develops. A gauge invariance results in the appearance of new particles. For a scalar theory, a Goldstone boson occurs. However, when a complex scalar theory is combined with a massless gauge field that forces the field to be real about an unbroken vacuum, the massive Higgs scalar field results. The gauge field itself also acquires mass.

As further developed in the model by Weinberg and Salam, the leptons and gauge bosons are combined into a single Lagrangian function that is based on a SU(2) × U(1) symmetry. Originally, all particles in the Lagrangian are massless, which include the massless electron neutrinos, electrons, and four gauge fields. However, as a consequence of spontaneous symmetry breaking, masses appear as a result of the Higgs mechanism when the SU(2) symmetry is broken. As such, the electron and the $W\pm$ and Z^0

particles that mediate the weak force now acquire mass. This process of symmetry breaking also gives rise to a photon field in the standard model thereby unifying the electromagnetic and weak interactions into a single Lagrangian.

As discussed in Section 4.2, neutrinos are experimentally observed to oscillate and switch identities challenging the long-held view that they are massless. Both the left-handed and right-handed neutrinos can acquire mass by interacting with the Higgs field. However, the right-handed neutrino can also interact with/couple to itself, thereby acquiring a so-called *Majorana* mass.

The individual Lagrangian formulations developed in the full standard model of Fig. D.1 are summarized below for the electromagnetic, weak and strong force. Recall that in these expressions, the generators of a given symmetry group in Fig. D.1 is a set of elements capable of producing all elements in the group. Thus, for the U(1) symmetry group, the number of generators is one. In general, for a SU(N) symmetry group, the number of generators is given as $N^2 - 1$.

D.3   Langrangian Functions of the Standard Model

Separate Lagrangian functions are needed for the electromagnetic, strong and weak forces. The Lagrangian functions for these three forces are listed below. The theoretical basis of these functions follows from quantum field theory as detailed in Appendix D.3.1.

(i) The symmetry group for quantum electrodynamics is the local U(1) Albelian (commutative) group. This Lagrangian pertains to a spin-1/2 field that interacts with the electromagnetic field. The overall form of this Lagrangian is

$$\mathcal{L}_{QED} = \bar{\psi}\left(i\gamma^{\mu}D_{\mu} - m\right)\psi - \frac{1}{4}F^{\mu\nu}F_{\mu\nu} \qquad (D.21)$$

In this equation, γ^{μ} are Dirac matrices from quantum mechanics, and ψ is a bispinor (two-component) field of the spin-1/2 particles (e.g., the electron-positron field). The parameter $D_{\mu} \equiv \partial_{\mu} - ieA_{\mu}$ is a "covariant derivative" of the gauge field, where e is a coupling constant, and A_{μ} is the potential of the electromagnetic field generated by the electron. In addition, m is the mass of the electron/positron and $F_{\mu\nu} = \partial_{\mu}A_{\nu} - \partial_{\nu}A_{\mu}$ is the electromagnetic field tensor used in the formulation of Maxwell's equations in Appendix A.2.

(ii) The interactions between quarks and gluons are described by an overall SU(3) symmetry with the fundamental generators being T_a. The leptons do not interact with gluons so that they are not affected by quantum chromodynamics. The Lagrangian function for the coupling of the quarks to the gluon fields is given by

$$\mathcal{L}_{QCD} = \sum_{\psi} \bar{\psi}_i \left(i\gamma^\mu \left[\partial_\mu \delta_{ij} - i g_s (T_a)_{ij} A_\mu^a \right] - m\delta_{ij} \right) \psi_j - \frac{1}{4} G_{\mu\nu}^a G_a^{\mu\nu}$$

(D.22)

In this expression, ψ_i is the Dirac spinor of the quark field where $i = \{\text{red, green, blue}\}$. Here γ^μ are the Dirac matrices, A_μ^a is the 8-component ($a = 1, 2, \ldots, 8$) SU(3) gauge gluon field and $(T_a)_{ij}$ are the 3×3 Gell–Mann matrices that are generators of the SU(3) colour group. The parameter $G_{\mu\nu}^a$ represents the gluon field strength tensor and g_s is the strong coupling constant.

(iii) The electroweak interaction obeys a U(1) × SU(2)$_L$ symmetry group. The subscript L indicates that coupling only occurs to the left-handed fermions in Fig. D.1. The Lagrangian function for the electroweak field is

$$\mathcal{L}_{EW} = \sum_{\psi} \bar{\psi}\gamma^\mu \left(i\partial_\mu - g'\frac{1}{2}Y_W B_\mu - g\frac{1}{2}\boldsymbol{\tau}_L \boldsymbol{W}_\mu \right) \psi$$

$$- \frac{1}{4} W_a^{\mu\nu} W_{\mu\nu}^a - \frac{1}{4} B^{\mu\nu} B_{\mu\nu}$$

(D.23)

In this expression, B_μ is the U(1) gauge field. The "weak hypercharge", Y_W, is a generator of the U(1) group that arises as a quantum number in the theory. It is related to the electric charge, Q, and the third component of weak isospin, T_3, through the relation $Q = T_3 + Y_W/2$ equivalent to the earlier Gell–Mann–Nishijima formula (or by the alternative convention $Q = T_3 + Y_W$ as shown in Fig. D.1). The parameter $\boldsymbol{W}_\mu$ is the three-component of the SU(2) gauge field. The components of $\boldsymbol{\tau}$ are the "Pauli matrices" derived from quantum mechanics, which are generators of the SU(2) symmetry group — the subscript L indicates that they only act on left-chiral fermions. The parameters g and g' are the U(1) and SU(2) coupling constants, respectively, in the weak theory. The parameter $W^{a\mu\nu}$ ($a = 1, 2, 3$) and $B^{\mu\nu}$ are the field strength tensors for the weak isospin and weak hypercharge fields. A new symmetry of U(1), which differs from that of quantum electrodynamics, is required in terms of the weak hypercharge for the unification of the weak force.

Table D.1. Weak quantum numbers of
leptons and quarks.

	T	T_3	Q	Y_W
Leptons				
ν_L	1/2	1/2	0	−1
e_L^-	1/2	−1/2	−1	−1
ν_R	0	0	0	0
e_R^-	0	0	−1	−2
Quarks				
u_L	1/2	1/2	2/3	1/3
d_L	1/2	−1/2	−1/3	1/3
u_R	0	0	2/3	4/3
d_R	0	0	−1/3	−2/3

The quantum numbers of the leptons and quarks in the electroweak
theory for the weak isospin and hypercharge are summarized in
Table D.1. This table provides values of the weak isospin (T),
third component of weak isospin (T_3), electric charge (Q) and weak
hypercharge (Y_W).

It should be noted that the right-handed neutrino ν_R does not
carry SU(2) × U(1) charges and therefore decouples from the weak
interaction.

D.3.1 *Quantum field theory fundamentals*

Quantum Electrodynamics Theory: Using the definition of the par-
tial derivative with respect to space–time where $\partial_\mu = \frac{\partial}{\partial x^\mu}$, with the time
component $\partial_0 = \frac{1}{c}\frac{\partial}{\partial t}$, the Dirac spinor field equation in Eq. (C.11) for a
free electron can be rewritten as

$$i\hbar\gamma^\mu\partial_\mu\psi(x) - mc\psi(x) = 0 \qquad (D.24)$$

or in natural units (where $\hbar$ and c equal unity):

$$(i\gamma^\mu\partial_\mu - m)\,\psi = 0 \qquad (D.25)$$

The gamma matrices γ^μ are a set of 4 × 4 complex matrices such that:
$\gamma^0 = \begin{pmatrix} I_2 & 0 \\ 0 & -I_2 \end{pmatrix}$ and $\gamma^i = \begin{pmatrix} 0 & \tau^i \\ -\tau^i & 0 \end{pmatrix}$, where I_2 is a 2 × 2 identity matrix and
τ^i are the Pauli matrices $\tau^1 = \begin{pmatrix} 0 & 1 \\ 1 & 0 \end{pmatrix}$, $\tau^2 = \begin{pmatrix} 0 & -i \\ i & 0 \end{pmatrix}$ and $\tau^3 = \begin{pmatrix} 1 & 0 \\ 0 & -1 \end{pmatrix}$. For
the chiral representation, γ^i are the same but $\gamma^0 = \begin{pmatrix} 0 & I_2 \\ I_2 & 0 \end{pmatrix}$.

The Dirac Lagrangian follows from Eq. (D.25) as:

$$\mathcal{L}_{\text{Dirac}}(\psi) = \bar{\psi}\left(i\gamma^{\mu}\partial_{\mu} - m\right)\psi.$$

To account for the coupling of the electromagnetic field to the electron, the ordinary space derivative ∂_{μ} must be replaced by the derivative $D_{\mu} = \partial_{\mu} - ieA_{\mu}$, where e is a coupling constant $\simeq 0.303$. Here $A_{\mu} = (\phi, \vec{A})$ is a four vector accounting for the electric potential ϕ and vector potential $\vec{A}$ for the electric field $\vec{E}$ and magnetic field $\vec{B}$, respectively. Thus, the coupled Lagrangians are

$$\mathcal{L}_{\text{Dirac}}(\psi) + \mathcal{L}_{\text{Coupling}}(\psi, A) = \bar{\psi}\left(i\gamma^{\mu}[\partial_{\mu} - ieA_{\mu}] - m\right)\psi. \qquad \text{(D.26)}$$

For instance, consider the coupling of an electron and photon, where such processes may include: (a) the absorption of a photon by an electron, (b) emission of a photon by an electron, (c) production of an electron–positron pair, and (d) annihilation of an electron–positron pair. By convention, the electron field ψ annihilates an electron. An anti-particle (i.e., positron) must be equally created in order to conserve charge. As such, the field ψ annihilates an electron and creates a photon. To create the electron, the conjugate field $\bar{\psi}$ must do the opposite with the annihilation of a positron and the creation of an electron. Since the photon is not electrically charged, the electromagnetic field A is capable of both annihilating and creating a photon. Thus, as an example, for the absorption of a photon by an electron in (a), one has the expression: $e\bar{\psi}(x)\gamma^{\mu}\psi(x)A_{\mu}(x)$. Reading from right to left, at the location x in space–time, the electromagnetic field A annihilates a photon, and the electron field ψ annihilates an electron. The conjugate field $\bar{\psi}$ then creates an electron. Here e is the electromagnetic coupling strength that fixes the amplitude for this process. Thus, overall, the electron absorbs the photon and at point x the photon disappears (i.e., is annihilated). But the electron is annihilated as well; however, this latter annihilation is immediately followed by the creation of an electron at the same point x.

Finally, the complete Lagrangian for quantum electrodynamics (QED) is

$$\mathcal{L}_{\text{QED}}(\psi, A) = \mathcal{L}_{\text{Dirac}}(\psi) + \mathcal{L}_{\text{Coupling}}(\psi, A) + \mathcal{L}_{\text{Maxwell}}(A)$$

This expression therefore describes how electrons and photons interact. It is obtained by combining the Lagrangian for Maxwell's equation in Example A.4 along with Eq. (D.26), yielding the final expression in Eq. (D.21).

Quantum Chromodynamics Theory: The derivation of the Lagrangian for quantum chromodynamics (QCD) follows the analogous development of QED. This treatment considers "Yang–Mills" theory that was initially developed to describe the strong force that binds nucleons (N) together in the nucleus. This theory was later advanced in accordance with "quark theory" to better reflect experimental observations. In this theory, the gluon is an elementary particle that acts as the exchange particle (or "gauge boson") for the strong force interaction to bind quarks together forming hadrons such as protons and neutrons.

In electromagnetics, one can shift A_μ to $A_\mu + \partial_\mu \Lambda$ without changing the electromagnetic field nor Maxwell's equations. Hence, one has the freedom to change A_μ to $A_\mu + \partial_\mu \Lambda$ with any function $\Lambda(x)$ of space–time without changing $\vec{E}$ and $\vec{B}$. In fact, Herman Weyl recognized that if $A_\mu \to A_\mu + \partial_\mu \Lambda$, then the electron field ψ also needs to transform with the use of a phase factor, $\psi \to e^{i\Lambda}\psi$, if Λ depends on x. In this way, changing $A_\mu(x)$ and $\psi(x)$ allows the Lagrangian $\mathcal{L}$ to remain unchanged (i.e., "gauge invariant").

To transfer the proton and neutron into one another, "isotropic spin" or so-called "isospin" provided a symmetry for the strong interaction. An isospin rotation acting on a nucleon field N for the protons p and neutrons n is specified by three angles θ^1, θ^2 and θ^3. Mathematically, for an isospin rotation, this gives rise to the field transformation $N \to e^{i(\theta^a T^a)}$, where $a = 1$, 2 and 3. Here T^a denotes three matrices acting on the two components of N. Heisenberg in fact recognized that the Lagrangian for the strong interaction is invariant under this "global transformation" because the θ numbers are three real numbers independent of x.

Comparing the two transformations for ψ and N, one notices that $\theta^a T^a$ is a two-by-two matrix, while $\Lambda(x)$ is a real number for electromagnetism. Since $\Lambda(x)$ is an arbitrary function of space–time, while $\theta^a T^a$ does not depend on position, there is no equivalent analogue of A_μ for the strong interaction. However, in 1954, Chen-ning Yang and Robert Mills wanted to force the two Lagrangians to look the same by making $\theta^a(x)$ a local quantity. In this case, one can then introduce three gauge fields A_μ^a that transform similarly as: $gA_\mu^a \to gA_\mu^a + \partial_\mu \theta^a$. Hence, the Yang–Mills theory yields three gauge fields A_μ^a instead of a single electromagnetic field A_μ. Each of these gauge fields is associated with a "Yang–Mills gauge boson particle" (just like the electromagnetic field is associated with a photon).

The Yang–Mills theory had problems because it dealt with the process of transforming a proton into a neutron. Consequently, this theory under-predicted the probability amplitude for the process as compared to experimental measurements by a factor of 3. Eventually Gell–Mann and others realized the need for a quark theory, where the transformation of a neutron into a proton and vice versa involves instead converting an up quark into a down quark and vice versa. Gell–Mann realized that quarks come in three copies, labelled by the three different colours of red, blue and green. As such, it was realized that the Yang–Mills theory should not be applied as an isospin transformation but to the transformation of these three colours. For example, in the scattering of two protons, the up and down quarks in the proton do not change into one another but rather the quarks change colour. The strong interaction is therefore based on an SU(3) symmetry acting on three colours. This requires $3^2 - 1 = 8$ gauge bosons rather than 3 envisaged in the original Yang–Mills model. The attribute that distinguishes an up quark from a down one is called flavour. Therefore, quarks come in six flavours (i.e., up, down, strange, charm, top and bottom) each in three colours as shown in Fig. D.1.

The interactions between quarks and gluons are therefore defined using a Yang–Mills gauge theory with an SU(3) symmetry that is generated by T_a. The leptons do not interact with gluons. Thus, the corresponding Dirac Lagrangian for the quarks coupled to the gluon fields, analogous to Eq. (D.26), is

$$\mathcal{L}_{\text{Dirac}} = \bar{\psi}_i \left(i\gamma^\mu (D_\mu)_{ij} - m\delta_{ij} \right) \psi_j.$$

Here $\psi_i(x)$ is the Dirac spinor of the quark field, which is a dynamical function of space–time as a SU(3) gauge group representation, where the index i corresponds to the three colours of red, blue and green. The index i and j run from 1 to 3 and δ_{ij} is the Kronecker delta. The parameters γ^μ are Dirac matrices. However, in this theory, the gauge covariant derivative instead takes the form: $(D_\mu)_{ij} = \partial_\mu \delta_{ij} - ig_s(T_a)_{ij}A_\mu^a$. This derivative couples the quark field to the gluon fields with a coupling strength g_s. $A_\mu^a(x)$ are the gluon fields that are functions of space–time where the indices a, b, c run from 1 to 8. The SU(3) generators $(T_a)_{ij}$ of the SU(3) colour group are 3×3 Gell–Mann matrices where $T_a = \lambda_a/2$ with $\lambda_a(a = 1, \ldots, 8)$:

$$\lambda_1 = \begin{pmatrix} 0 & 1 & 0 \\ 1 & 0 & 0 \\ 0 & 0 & 0 \end{pmatrix}, \quad \lambda_2 = \begin{pmatrix} 0 & -i & 0 \\ i & 0 & 0 \\ 0 & 0 & 0 \end{pmatrix}, \quad \lambda_3 = \begin{pmatrix} 1 & 0 & 0 \\ 0 & -1 & 0 \\ 0 & 0 & 0 \end{pmatrix},$$

$$\lambda_4 = \begin{pmatrix} 0 & 0 & 1 \\ 0 & 0 & 0 \\ 1 & 0 & 0 \end{pmatrix}, \quad \lambda_5 = \begin{pmatrix} 0 & 0 & -i \\ 0 & 0 & 0 \\ i & 0 & 0 \end{pmatrix}, \quad \lambda_6 = \begin{pmatrix} 0 & 0 & 0 \\ 0 & 0 & 1 \\ 0 & 1 & 0 \end{pmatrix},$$

$$\lambda_7 = \begin{pmatrix} 0 & 0 & 0 \\ 0 & 0 & -i \\ 0 & i & 0 \end{pmatrix}, \quad \lambda_8 = \frac{1}{\sqrt{3}} \begin{pmatrix} 1 & 0 & 0 \\ 0 & 1 & 0 \\ 0 & 0 & -2 \end{pmatrix}.$$

The variable m corresponds to the quark mass, where the parameters m and g_s are subject to renormalization.

Similar to the Maxwell field contribution in Eq. (D.21) involving the electromagnetic field tensor strength, $F^{\mu\nu}$, an analogous component to the Lagrangian density can be added for the gluon field strength. In this case, one replaces $F^{\mu\nu}$ by the tensor $G^a_{\mu\nu}$, where

$$G^a_{\mu\nu} = \partial_\mu A^a_\nu - \partial_\nu A^a_\mu + g_s f^{abc} A^b_\mu A^c_\nu$$

and f^{abc} are structure constants of the SU(3) group. Thus, the total Lagrangian density can be described by Eq. (D.22).

Electroweak Theory: The electroweak theory provides unification of the weak and electromagnetic interactions between elementary particles. This work led to a Noble Prize for Sheldon Glashow, Abdus Salam and Steven Weinberg in 1979. The existence of the electroweak interaction was proven experimentally in neutrino scattering experiments with the discovery of "neutral current" interactions, where subatomic particles interact by means of the weak force with the mediation of the uncharged Z boson. This work subsequently led to the discovery of the W and Z boson in 1999 in the Super Proton Synchrotron.

The weak and electroweak interactions have unique properties. The range of the weak interaction is some 600 times shorter than that of the strong interaction. This means that the particle mediating the weak interaction, i.e., the W boson, is about 600 times more massive than the meson as proposed by Yukawa for the strong interaction. In the weak interaction, for example, in a more elementary process, the neutron inside the nucleus transmutes itself into a proton while emitting an electron and anti-neutrino: $n \rightarrow p + e^- + \bar{\nu}$. The proton is made up of two up quarks and a down quark as $P = (uud)$, while the neutron is made up of two down quarks and an up quark as $N = (udd)$. Here, the d quark transforms into a u quark by emitting a virtual exchange W^- particle, which subsequently morphs into an e^- and an anti-neutrino $\bar{\nu}$.

Interestingly, in the fundamental laws of Nature, the weak interaction breaks parity symmetry (P) as a chiral process with a preference for right- or left-handedness. Parity violation can best be explained in accordance with the helicity for the spin of a moving particle. The neutrino was in fact discovered to be left-handed. The weak interaction also violates charge conjugation invariance (or C) for the changing of matter into anti-matter and vice-versa. The laws of physics normally respect time reversal (T). However, in the weak decay of the K meson, it was discovered that T can also be violated, although the combination of CPT is never violated.

Given parity violation, the field ψ that Dirac used to describe the electron (and later used by Gell–Mann to describe quarks) may be split into a left-handed and right-handed field: $\psi = \psi_L + \psi_R$. Thus, parity violation in the weak interaction is accommodated for by ignoring the right-handed fields for the W bosons.

Another peculiar nature of particle physics is that although the gauge bosons, as mediators of the three fundamental interactions (i.e., the photon of the electromagnetic interaction, W and Z bosons of the weak interaction, as well as the gluons of the strong interaction), are not repeated as separate families. What is repeated three times are the quarks and leptons (with their associated neutrinos) for the matter content of the universe as depicted in Fig. D.1. Moreover, a coupling or "mixing" between the down, strange and bottom quarks has been experimentally observed between families for the transmutation of quarks in the weak decay process. Here the weak interaction eigenstates of quarks are not identical to their mass eigenstates. This leads to quarks changing from one type of flavour into another via weak decays. This process is described by the Cabibbo-Kobayashi-Maskawa (CKM) matrix. This introduces CP violation in the Standard Model, which is important to explain matter-antimatter asymmetry in the universe.

As previously shown, the Yang–Mills theory must be intricately balanced for gauge invariance to hold. The Higgs mechanism is able to generate masses, for example, for the W bosons without affecting the gauge invariance of the Lagrangian. The number of gauge bosons in a Yang–Mills theory is specifically fixed by group theory. It was initially thought that for the SU(2) symmetry, the three gauge bosons were W^+, W^- and the photon. However, it was eventually shown that a fourth gauge Z boson is needed. As such, the group structure has to be SU(2) × U(1).

The Lagrangian function for the electroweak formalism before symmetry breaking consists of the following components:

$$\mathcal{L}_{\text{EW}} = \mathcal{L}_{\text{gauge}} + \mathcal{L}_{\text{matter}} + \mathcal{L}_{\text{Higgs}} + \mathcal{L}_{\text{Yukawa}} \qquad (D.27)$$

The $\mathcal{L}_{\text{gauge}}$ term describes the interaction between the three W vector bosons and the B vector boson:

$$\mathcal{L}_{\text{gauge}} = -\frac{1}{4}W_a^{\mu\nu}W_{\mu\nu}^a - \frac{1}{4}B^{\mu\nu}B_{\mu\nu} \tag{D.28}$$

Here $W_{\mu\nu}^a$ ($a = 1, 2, 3$) and $B_{\mu\nu}$ are the field strength tensors for the weak isospin and weak hypercharge gauge fields given by

$$W_{\mu\nu}^i = \partial_\mu W_\nu^i - \partial_\nu W_\mu^i - ig\epsilon^{ijk}W_\mu^j W_\nu^k$$

$$B_{\mu\nu} = \partial_\mu B_\nu - \partial_\nu B_\mu$$

In this expression, B_μ is the U(1) gauge field, W_μ^i corresponds to the SU(2) gauge field components and ϵ^{ijk} is the Levi-Civita symbol.

The second component of the electroweak Lagrangian, $\mathcal{L}_{\text{matter}}$, is the kinetic term for the fermions in the Standard Model:

$$\mathcal{L}_{\text{matter}} = \bar{Q}_j i\gamma^\mu D_\mu Q_j + \bar{u}_j i\gamma^\mu D_\mu u_j + \bar{d}_j i\gamma^\mu D_\mu d_j + \bar{L}_j i\gamma^\mu D_\mu L_j$$

$$+ \bar{e}_j i\gamma^\mu D_\mu e_j \tag{D.29}$$

where the subscript j sums over the three generations of fermions. Here Q, u, d correspond to the left-handed doublet, right-handed singlet up, and right-handed singlet down quark fields (see Fig. D.1). The parameters L and e are the left-handed doublet and right-handed singlet electron fields. The gauge bosons and fermions interact through the gauge covariant derivative (excluding the gluon gauge field for the strong interaction):

$$D_\mu = \partial_\mu + ig\prime\frac{1}{2}Y_W B_\mu + ig\frac{1}{2}T_j W_\mu^j \tag{D.30}$$

where Y_W is the weak hypercharge and T_j are the components of the weak isospin. The parameter g and $g\prime$ are the U(1) and SU(2) coupling constants. The last two equations, Eqs. (D.29) and (D.30), can be more succinctly written for the Lagrangian function of the electroweak field as

$$\mathcal{L}_{EW} = \sum_L \bar{L}\gamma^\mu \left(i\partial_\mu - g\prime\frac{1}{2}Y_W B_\mu - g\frac{1}{2}\vec{\tau}\cdot\vec{W}_\mu \right) L$$

$$+ \sum_R \bar{R}\gamma^\mu \left(i\partial_\mu - g\frac{1}{2}\vec{\tau}\cdot\vec{W}_\mu \right) R$$

$$= \sum_\psi \bar{\psi}\gamma^\mu \left(i\partial_\mu - g\prime\frac{1}{2}Y_W B_\mu - g\frac{1}{2}\tau_L W_\mu \right) \psi \tag{D.31}$$

where L denotes a left-handed fermion doublet and R is a right-handed fermion singlet. The parameter $\boldsymbol{W}_\mu$ is the three-component of the SU(2) gauge field. The components of $\boldsymbol{\tau}$ are the "Pauli matrices" where the subscript L again indicates that they only act on left-chiral fermions.

The third term of the electroweak Lagrangian, $\mathcal{L}_{\text{Higgs}}$, describes the Higgs field h and its interactions with itself and the gauge bosons:

$$\mathcal{L}_{\text{Higgs}} = |D_\mu h|^2 - \lambda \left(|h|^2 - \frac{v^2}{2} \right)^2 \tag{D.32}$$

The parameter v is the vacuum expectation value and λ a constant for the Higgs potential $V(\phi) = -\mu^2 \phi^\dagger \phi + \lambda (\phi^\dagger \phi)^2$ as described in Appendix D.1.

Finally the fourth term in the Lagrangian, $\mathcal{L}_{\text{Yukawa}}$, accounts for the interaction between a scalar field ϕ and a Dirac field for the fermions ψ:

$$\begin{aligned}
\mathcal{L}_{\text{Yukawa}}(\phi, \psi) &= -g^f \bar{\psi} \phi \psi \\
&= -\sum_{f-} g^f_- \left(\bar{L} \phi R + \bar{R} \bar{\phi} L \right) - \sum_{f+} g^f_+ \left(\bar{L} \phi^c R + \bar{R} \bar{\phi}^c L \right)
\end{aligned} \tag{D.33}$$

where $g^f_\pm$ is the fermion Yukawa coupling constant for $T_3 = \pm \frac{1}{2}$. Here L denotes a left-handed fermion doublet and R is a right-handed fermion singlet. The ϕ^c indicates a charge conjugation of the Higgs doublet where $\phi^c = i\tau_2 \phi^\dagger$. This contribution will generate masses of the fermions when the Higgs field acquires a non-zero expectation value.

The Higgs field is a scalar field (spin 0) which is a two-component object. After symmetry breaking one has

$$\phi(x) = \frac{1}{\sqrt{2}} \begin{pmatrix} 0 \\ v + h(x) \end{pmatrix}$$

Masses are acquired through the Higgs mechanism as demonstrated in Appendix D.2. In particular, the W bosons acquire a mass of $m_W = 2gv = 80.4$ MeV. The mass of the Z boson $m_Z = \frac{m_W}{cos\theta_W}$, where θ_W is the Weinberg angle, yielding a value of 91.19 GeV for this particle. The fermions acquire mass as $m_f = \frac{g^f v}{\sqrt{2}}$. In particular, such masses include: the electron mass $m_e = 511$ keV, muon mass $m_\mu = 105.7$ MeV, tau mass $m_\tau = 1.78$ GeV, up quark mass $m_u = 1.9$ MeV, down quark mass $m_d = 4.4$ MeV, strange quark mass $m_s = 87$ MeV, charm quark mass $m_c = 1.32$ MeV, bottom quark mass $m_b = 4.24$ MeV, and top quark $m_t = 173.5$ GeV. The boson particle has a mass of $m_h = v\sqrt{2\lambda} = 125$ GeV (as determined in the Large

Electron-Proton Collider (LEP)). The Higgs vacuum expectation value is $v = 246$ GeV. The U(1), SU(2) and SU(3) gauge coupling constants are $g' = 0.357$, $g = 0.652$ and $g_s = 1.221$, respectively.

The masses in the general formula of Eq. (D.23) account for the gauge and fermion components depicted in Eqs. (D.28) and (D.31).

E   String Theory

A calculation is performed for a simple bosonic string. This analysis shows key aspects of string theory. It reveals how the critical (extended) dimensions of space–time arise in the theory (Appendix E.1) and how these extended dimensions can compactify (Appendix E.2).

E.1   Extended Dimensions for an Open Boson String

As an illustrative example, consider the simpler case of an open boson string (Type I). This analysis provides for a determination of the critical number of space–time dimensions needed to avoid an unphysical state with string quantization. For the quantization, commutation relations are needed. These relations have a similarity to the harmonic oscillator encountered in quantum physics in Appendix C. For the harmonic oscillator, one has creation and annihilation operators $\hat{a}^\dagger$ and $\hat{a}$ such that

$$\hat{a}^\dagger = \sqrt{\frac{m\omega}{2\hbar}}\left(\hat{x} - \frac{i}{m\omega}\hat{p}\right), \text{ and } \hat{a} = \sqrt{\frac{m\omega}{2\hbar}}\left(\hat{x} + \frac{i}{m\omega}\hat{p}\right)$$

with the commutation relation $[\hat{a}, \hat{a}^\dagger] = 1$. The Hamiltonian for the oscillator system is

$$\hat{H} = \hbar\omega\left(\hat{a}^\dagger\hat{a} + \frac{1}{2}\right)$$

The number operator $\hat{N} = \hat{a}^\dagger\hat{a}$ has the eigenstates $|n\rangle$ where

$$\hat{N}|n\rangle = n|n\rangle, \quad \text{where } n = 0, 1, 2, \ldots.$$

Thus, the quantized energy levels are

$$\hat{H}|n\rangle = \hbar\omega\left(\hat{N} + \frac{1}{2}\right)|n\rangle = \hbar\omega\left(n + \frac{1}{2}\right)|n\rangle = E_n|n\rangle$$

The ground state of the system is $|0\rangle$.

One has a similar system for the vibrating string with an analogous commutation relation. Here $[\alpha_m^\nu, \alpha_n^\nu] = \eta^{\mu\nu} m \delta_{m+n,0}$, which yields

$$[\alpha_m^\nu, \alpha_n^\nu] = [\alpha_m^\nu, \alpha_{-m}^\nu] = m\eta^{\mu\nu} \tag{E.1}$$

The presence of the Minkowski metric $\eta^{\mu\nu}$ can result in a negative commutator since $\eta^{00} = -1$:

$$[\alpha_m^0, \alpha_{-m}^0] = -m \tag{E.2}$$

Analogous to the harmonic oscillator, an equivalent number operator is

$$N_m = \alpha_{-m} \cdot \alpha_m \quad \text{for } m \geq 1. \tag{E.3}$$

The eigenstates of the number operator similarly satisfy:

$$N_m |i_m\rangle = i_m |i_m\rangle \tag{E.4}$$

The total number operator is obtained by summing over the values of m

$$N = \sum_{m=1}^{\infty} N_m = \sum_{m=1}^{\infty} \alpha_{-m} \cdot \alpha_m \tag{E.5}$$

Analogous to quantum mechanics, α_m^μ acts like a lowering operator while $\alpha_{-m}^\mu, (m \geq 1)$ acts like a raising operator

$$N_m = (\alpha_{-m}|i_m\rangle) = (i_m + m)(\alpha_{-m}|i_m\rangle) \tag{E.6}$$

The lowering operator α_m^μ destroys the vacuum state (i.e., ground state) $i_m = 0$

$$\alpha_m^\mu |0\rangle = 0 \tag{E.7}$$

Consider the ground state where the string carries momentum k^μ. The momentum operator acts as

$$p^\mu |0, k\rangle = k^\mu |0, k\rangle \tag{E.8}$$

Thus, the "norm" for the first excited state is evaluated using $(\alpha_{-1}^0)^\dagger = \alpha_1^0$:

$$|\alpha_{-1}^0|0, k\rangle| = \langle 0, k|\alpha_1^0 \alpha_{-1}^0|0, k\rangle = -1 \tag{E.9}$$

This result indicates a negative norm state which is unphysical. This arises because of the negative time component in the metric. However, one can rid the theory of negative norm states by applying so-called "Virasoro constraints". The classical expression for these constraints is

$$L_m = \frac{1}{2} \sum_n \alpha_{m-n} \cdot \alpha_n \tag{E.10}$$

For the string analysis, this expression is replaced by a "Virasoro operator" where the "normal ordering" of states must be followed:

$$L_m = \frac{1}{2} \sum_n : \alpha_{m-n} \cdot \alpha_n : \tag{E.11}$$

The colons indicate that all lowering operators move to the right and all raising operators to the left. Ordering guarantees that the resulting eigenfunctions are finite. The operator quantity L_m lowers the eigenvalue of the operator by m. From the commutation relation in Eq. (E.1), the operators a_{m-n}^μ and a_n^μ will commute when $m \neq 0$ so that one can simply move the raising and lowering operators around in the Virasoro operator as no extra terms arise from the commutator. For a normal-ordered result:

$$L_0 = \frac{1}{2}\alpha_0^2 + \sum_{n=1}^\infty \alpha_{-n} \cdot \alpha_n \tag{E.12}$$

Note that for the more general expression for $m = 0$, using the commutator relation in Eq. (E.1) for the modes, one obtains:

$$\frac{1}{2}\sum_{n=-\infty}^{n=\infty} \alpha_{-n} \cdot \alpha_n = \frac{1}{2}(\alpha_0 \cdot \alpha_0) + \frac{1}{2}\sum_{n=1}^\infty \alpha_{-n} \cdot \alpha_n + \frac{1}{2}\sum_{n=1}^\infty \alpha_n \cdot \alpha_{-n}$$

$$= \frac{1}{2}(\alpha_0 \cdot \alpha_0) + \sum_{n=1}^\infty \alpha_{-n} \cdot \alpha_n + \frac{1}{2}\sum_{\mu=0}^{D-2} \eta_\mu^\mu \sum_{n=1}^\infty n$$

$$= \frac{1}{2}(\alpha_0)^2 + \sum_{n=1}^\infty \alpha_{-n} \cdot \alpha_n + \frac{D-2}{2}\sum_{n=1}^\infty n \tag{E.13}$$

The last term in the third line of Eq. (E.13) is omitted in Eq. (E.12) with ordering. Although this last term appears infinite, one can use a "regularization" to compute a finite value for this single term. The Riemann-zeta function can be employed in this case where

$$\zeta(s) = \sum_{n=1}^\infty n^{-s} \tag{E.14}$$

It is noted that the sum in the last term of Eq. (E.13) corresponds to $s = -1$. Using the unique analytic continuation property of the Riemann-zeta function [Spiegel, 1973]:

$$\zeta(1 - x) = 2^{1-x}\pi^{-x}\Gamma(x)\cos\left(\frac{\pi x}{2}\right)\zeta(x) \qquad (E.15)$$

Hence, for $x = 2$, the function $\zeta(-1) = -1/12$ since the gamma function $\Gamma(2) = 1$, the Riemann-zeta function $\zeta(2) = \frac{\pi^2}{6}$ and $\cos(\pi) = -1$. Thus, it follows that

$$\frac{D-2}{2}\sum_{n=1}^{\infty} n = -\frac{D-2}{24} \qquad (E.16)$$

Of further interest is the difference between the general expression of L_0 in Eq. (E.13) and the normal ordered expression of L_0 in Eq. (E.12). This difference is denoted by a constant a so that for any expression involving the calculation of L_0, L_0 is replaced by $L_0 - a$.

Using the commutation relation in Eq. (E.1), the Virasoro operator commutes according to a "*Virasoro algebra with central extension*":

$$[L_m, L_n] = (m - n)L_{m+n} + \frac{D}{12}(m^3 - m)\delta_{m+n,0} \qquad (E.17)$$

The central charge is the space–time dimension D in the second term on the right-hand side of Eq. (E.17), which indicates the number of free scalar fields. However, if $m = 0, \pm 1$, the central term vanishes providing a closed algebra in the form of a "(2,R) algebra" for the single parameters L_1, L_2 and L_{-1}.

Importantly, the Virasoro operators provide a way to eliminate unphysical states (i.e., negative norm states). This procedure is performed by requiring that the expectation value for $L_0 - a$ vanishes for a physical state $|\psi\rangle$ with the condition:

$$\langle\psi|L_m - a\delta_{m,0}|\psi\rangle = 0 \quad \text{for } m \geq 0. \qquad (E.18)$$

The term $a\delta_{m,0}$ takes care that the normal ordering constant a is only needed for L_0. Thus, to eliminate unphysical states, one specifies conditions for a and D. The parameter D specifically corresponds to the extra space–time dimensions in the theory. In fact, the negative norm states can be eliminated if $a = 1$ and $D = 26$ as shown in the following analysis.

A mass operator follows from Einstein's special theory of relativity where: $p^\mu p_\mu + M^2 = 0 \Rightarrow M^2 = -p_\mu p^\mu$. As such, the first term $\frac{1}{2}\alpha_0^2$ in

Eq. (E.12) is associated with mass, i.e.,

$$\frac{1}{2}\alpha_0^2 = -\alpha' M^2 \tag{E.19}$$

The parameter α' is related to the tension of the string T with $\alpha' = 1/(2\pi T)$. The term $\sqrt{2\pi T}$ sets the energy scale on the order of the "Planck mass". Hence, the mass operator follows from the condition that with $a = 1$:

$$(L_0 - a)|\psi\rangle = 0 \Rightarrow \left(\frac{1}{2}\alpha_0^2 + \sum_{n=1}^{\infty} \alpha_{-n} \cdot \alpha_n - 1 \right)|\psi\rangle = 0$$

$$\Rightarrow -\alpha' M^2 + N - 1 = 0$$

$$\Rightarrow M^2 = \frac{1}{\alpha'}(N - 1) \tag{E.20}$$

where the total number operator N in Eq. (E.5) has been used. The third line in Eq. (E.20) is termed the "mass shell" condition. As equivalently follows from the last line of Eq. (E.13) using Eqs. (E.16) and (E.19), the mass condition in Eq. (E.20) can be further expressed as:

$$M^2 = \frac{1}{\alpha'}\left(N - \frac{D - 2}{24} \right) \tag{E.21}$$

On equating Eq. (E.21) to Eq. (E.20), this condition explicitly forces $D = 26$ *dimensions.*

Furthermore, the number operator acts on the ground state as

$$N|0\rangle = 0 \tag{E.22}$$

Similarly, using the mass operator in Eqs. (E.21) and (E.22) with $D = 26$, the mass of the ground state is

$$M^2|0\rangle = \frac{1}{\alpha'}\left(N - \frac{D - 2}{24} \right)|0\rangle = -\frac{1}{\alpha'}\frac{D - 2}{24}|0\rangle = -\frac{1}{\alpha'}|0\rangle \tag{E.23}$$

Hence, the mass of the ground state for an open boson string is *negative.* This result indicates that the ground state is an unphysical "tachyon" in this case, which cannot be eliminated in the theory. Consequently, one requires supersymmetry to remove the Tachyon for a more realistic theory.

One can further evaluate the mass of the first excited state. The first excited state is $|i\rangle = \alpha_{-1}^i|0\rangle$ where $i = 1, 2, \ldots, D - 2$. A vector with a spin-1 state has $D - 1$ components, while a state with $D - 2$ components is a massless state. This explains why the summation range in Eq. (E.13) goes up to $D - 2$. The mass of the excited state is

$$M^2 \alpha^i_{-1}|0\rangle = \frac{1}{\alpha'}\left(1 - \frac{D-2}{24}\right)\alpha^i_{-1}|0\rangle = -\frac{1}{\alpha'}\frac{26 - D}{24}\alpha^i_{-1}|0\rangle \qquad (E.24)$$

Thus, for the state to be massless, the term $(26 - D)/24$ must vanish, indicating again that the "critical dimension" for the open boson string is 26 dimensions.

E.2 Compactification in a Closed Boson String

Compactification enables the shrinkage of extra dimensions into a circle of radius R. Moreover, a "T-duality" provides for relatable theories. In this correspondence, the extra dimension that is compacted into a radius R is related to an extra dimension being compacted to a radius α'/R.

For simplicity, consider a closed bosonic string of dimension 26. The X^0 dimension is time-like, while the remaining ones are spatial dimensions. Consider a single spatial dimension X^{25} that is curled up into a circle of radius R. Since the ends of a closed string are joined, there is a "periodic boundary condition" for the string:

$$X^\mu(\sigma, \tau) = X^\mu(\sigma + 2\pi, \tau) \qquad (E.25)$$

where σ refers to spatial coordinates and τ to a temporal one. This boundary condition corresponds to a string moving in space–time with non-compact dimensions. However, consider the 25th-dimension to be compacted as a circle so that the boundary condition for X^{25} changes to

$$X^\mu(\sigma + 2\pi, \tau) = X^\mu(\sigma, \tau) + 2\pi n R \qquad (E.26)$$

This string now has "winding states", where the string can wind around the compacted dimension any number of times. For all of the other dimensions, Eq. (E.25) holds. In Eq. (E.26), the number n is called the "winding number". The winding w can then be defined as

$$w = \frac{nR}{\alpha'} \qquad (E.27)$$

Here the winding acts as a type of momentum.

F   Cosmology

The "Friedmann equation" in Appendix F.1 is of particular importance in the field of cosmology. It describes the expansion of the universe from the occurrence of the Big Bang, including its topological properties (i.e., curvature). In addition to the Friedmann equation, one requires a knowledge of how the density ρ as well as the pressure p of material evolve with time in the universe as described by the "fluid equation" in Appendix F.2. The "acceleration equation" in Appendix F.3 describes the rate of expansion of the universe. These cosmological equations are embodied in general relativity. The physical topology and evolution of the universe with time is detailed in Appendix F.4. This section describes "Hubble's law" and the evolution of the universe from predicted cosmological models. Astronomical observations for evidence of the expansion and accelerating nature of the universe is also presented. Finally, "inflation theory" is discussed that provides an explanation for several anomalies in the hot big bang model.

F.1   Friedmann Model

Consider a spherical volume $\frac{4\pi r^3}{3}$ of density ρ. From Newton's law of gravity, the gravitational force outside of a spherical object of unknown density depends only on its total mass M as if the mass was concentrated at a central point within the sphere. Thus, the gravitational potential energy V from Newton's law of gravity for a particle of mass m that is at a distance r away from a total mass $M = \frac{4\pi r^3}{3}\rho$ is

$$V = -\frac{GMm}{r} = -\frac{4\pi G\rho r^2 m}{3} \tag{F.1}$$

where G is the universal gravitational constant. It is recognized that gravity always attracts. The kinetic energy T for this test particle with velocity $\dot{r}$ (where the dot represents the time derivative of r) is

$$T = \frac{1}{2}m\dot{r}^2 \tag{F.2}$$

Hence, the total energy from the conservation of energy using Eqs. (F.1) and (F.2) is

$$U = \frac{1}{2}m\dot{r}^2 - \frac{4\pi G\rho r^2 m}{3} \tag{F.3}$$

Since the universe is homogeneous and expanding in a uniform fashion, a "comoving coordinate system" can be considered. This coordinate system is carried along with the expansion, where the real distance $\vec{r}$ is proportional to the comoving distance $\vec{x}$. The proportionality constant itself only depends on time and is known as the "scale factor" of the universe:

$$\vec{r} = a(t)\,\vec{x} \tag{F.4}$$

Substituting Eq. (F.4) into Eq. (F.3) gives

$$U = \frac{1}{2}m\dot{a}^2 x^2 - \frac{4\pi G\rho a^2 x^2 m}{3} \tag{F.5}$$

In this derivation, it is recognized that for comoving coordinates, $\dot{x} = 0$ since objects are fixed in this frame of reference. Rearranging Eq. (F.5) yields:

$$\boxed{\left(\frac{\dot{a}}{a}\right)^2 = \frac{8\pi G\rho}{3} - \frac{kc^2}{a^2}} \tag{F.6}$$

where $k = -2U/(mx^2c^2)$ is a constant. The parameter k must be a constant in order to guarantee homogeneity of the universe. As such, $U \propto x^2$ at different separation distances. The Friedmann equation specifically applies to large scales where distant galaxies are expanding farther apart. This expansion is like that of an inflating balloon that pulls galaxies along with its expansion. The relation in Eq. (F.6) governs the time evolution of the scale factor $a(t)$.

F.2 Fluid Equation

From the laws of thermodynamics, a change in energy dE along with a pressure p and a change in volume dV, for a unit comoving radius in

Eq. (F.4), relates to a change in entropy dS at temperature T:

$$dE + pdV = TdS \tag{F.7}$$

Thus, at a physical radius a as follows from Eq. (F.4), the energy within the spherical volume, in accordance with Einstein's equation $E = mc^2$, is

$$E = \frac{4\pi a^3}{3} \rho c^2 \tag{F.8}$$

Calculating the rate of change of the energy with respect to time is

$$\frac{dE}{dt} = 4\pi a^2 \rho c^2 \frac{da}{dt} + \frac{4\pi a^3}{3} \frac{d\rho}{dt} c^2 \tag{F.9}$$

Similarly, the rate of change of volume with respect to time is

$$\frac{dV}{dt} = 4\pi a^2 \frac{da}{dt} \tag{F.10}$$

Considering an adiabatic process where $dS = 0$, using Eqs. (F.9) and (F.10), one obtains the so-called "fluid equation":

$$\boxed{\dot{\rho} + 3\frac{\dot{a}}{a}\left(\rho + \frac{p}{c^2}\right) \doteq 0} \tag{F.11}$$

Here the dots again refer to time derivatives. This equation governs the time evolution of the mass density $\rho(t)$.

F.3   Acceleration Equation

By differentiating Eq. (F.6) with respect to time, substituting in $\dot{\rho}$ from Eq. (F.11), and using Eq. (F.6) again, gives the "acceleration equation" for the cosmological model of the universe:

$$\boxed{\frac{\ddot{a}}{a} = -\frac{4\pi G}{3}\left(\rho + \frac{3p}{c^2}\right)} \tag{F.12}$$

Derivation of Cosmological Equations from General Relativity: These same cosmological equations can also be developed directly from the general field equations of relativity. The "metric" in general relativity can be used for interpretation of the geometry of the universe. Consider the distance Δs between two comoving points x_1 and x_2:

$$\Delta s^2 = a(t) \left[\Delta x_1^2 + \Delta x_2^2\right]$$

where $a(t)$ is the scale factor of the universe. In four-dimensional space–time, the distance can be more generally written as

$$ds^2 = \sum_{\mu,\nu} g_{\mu\nu} dx^\mu dx^\nu$$

where $g_{\mu\nu}$ is the metric and μ and ν take on values of 0, 1, 2, and 3 such that x^0 is the time coordinate and the three other values of ν correspond to the spatial coordinates x^1, x^2 and x^3. The cosmological principle and homogeneity of the universe implies that the universe has constant curvature. The most general spatial metric in spherical polar coordinates is

$$ds_3^2 = \frac{dr^2}{1 - kr^2} + r^2 \left(d\theta^2 + \sin^2\theta d\phi^2\right)$$

Here k reflects the geometry of space with it being positive, zero or negative (see the next section). Since space can grow or shrink with time, this metric can be further generalized as the "Robertson–Walker metric":

$$ds^2 = -c^2 dt^2 + a^2(t) \left[\frac{dr^2}{1 - kr^2} + r^2 \left(d\theta^2 + \sin^2\theta d\phi^2\right)\right]$$

The metric is part of Einstein's field equations:

$$R_\nu^\mu - \frac{1}{2} g_\nu^\mu R = \frac{8\pi G}{c^4} T_\nu^\mu$$

where T_ν^μ is the energy–momentum tensor of matter, and R_ν^μ and R are the Ricci tensor and scalar, respectively. The latter quantities account for the curvature of space–time. There are ten equations considering the symmetry of the tensors.

Considering a perfect fluid with no viscosity or heat flow, the energy-momentum tensor can be written as

$$T_\nu^\mu = diag(-\rho c^2, p, p, p)$$

where ρ is the mass density and p the pressure. There are two independent Einstein equations, the time-time one, and the space–space one. The time-time one produces the Friedmann equation in Eq. (F.6). The second equation gives

$$2\frac{\ddot{a}}{a} + \left(\frac{\dot{a}}{a}\right)^2 + \frac{kc^2}{a^2} = -8\pi G \frac{p}{c^2}$$

Subtracting the Friedmann equation from this equation yields the acceleration equation in Eq. (F.12). Reversing the procedure from the way that the acceleration equation is derived from the Friedmann equation and fluid equation, yields the fluid equation in Eq. (F.11).

The fluid equation can alternatively be developed directly from the energy–momentum tensor that automatically includes energy conservation, where $T^\mu_{\nu;\mu} = 0$. The semi-colon corresponds to a covariant derivative and a summation occurs for the repeated index μ in accordance with the Einstein convention. The covariant derivative can be described in terms of the Christoffel symbol that depends on the metric such that

$$\Gamma^\nu_{\delta\alpha} = \frac{1}{2}g^{\mu\nu}[g_{\mu\alpha,\delta} + g_{\delta\mu,\alpha} - g_{\alpha\delta,\mu}]$$

and $g_{ij,k} \equiv \frac{\partial g_{ij}}{\partial x_k}$. An equivalent nomenclature for $g_{ij,k}$, as also used in the literature and in Eq. (B.11), is $g_{ij/k}$ with the comma replaced by a slash to indicate differentiation. The covariant derivative is defined as

$$T^\mu_{\nu;\mu} = T^\mu_{\nu,\mu} + \Gamma^\mu_{\alpha\mu}T^\alpha_\nu - \Gamma^\alpha_{\nu\mu}T^\mu_\alpha = 0$$

where the comma corresponds to a normal derivative. For the $\nu = 0$ component, since T^μ_ν is diagonal, the Christoffel symbols are: $\Gamma^0_{00} = 0$ and $\Gamma^1_{01} = \Gamma^2_{02} = \Gamma^3_{03} = \frac{\dot{a}}{a}$. Thus, substituting in these parameters and summing over repeated indices gives the same result for the fluid equation in Eq. (F.11).

F.4   Geometry and Evolution of the Universe

The value of k in Eq. (F.6) dictates the type of geometry that the universe will exhibit (see Fig. F.1). In particular, the universe will have either a "flat", "closed" or "open" structure as shown in Fig. F.2. The simplest type of geometry is a plane (i.e., flat geometry) where $k = 0$. On the other hand, a spherical geometry results with $k > 0$. This geometry is referred to as a closed universe since it has a finite size. The final choice is $k < 0$. This choice yields a hyperbolic geometry that looks like a saddle. This type of geometry is known as an open universe since parallel lines never meet indicating that the universe is infinite in extent.

F.4.1   Hubble's law

In accordance with the Hubble law, galaxies are moving away (i.e., receding) from the Earth at speeds proportional to their distance, in which the

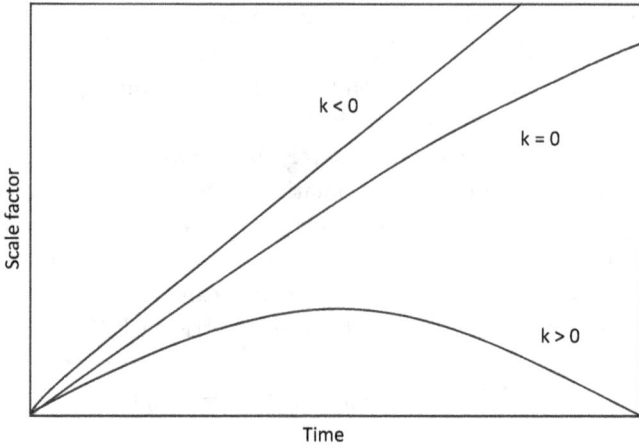

Figure F.1. Possible evolutions of the universe.

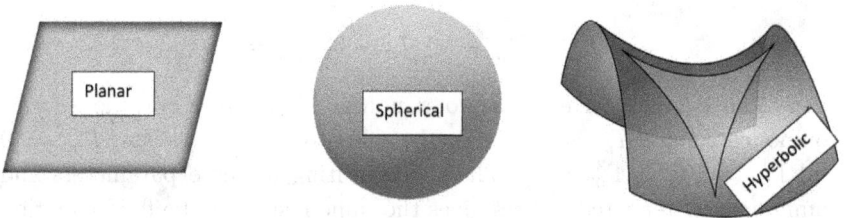

Figure F.2. Three different geometries for the universe showing a flat geometry (planar), closed geometry (spherical) and open geometry (hyperbolic).

velocities of the galaxies are estimated by their "red-shift". The velocity of recession is given by

$$\vec{v} = \frac{d\vec{r}}{dt} = \frac{|\dot{\vec{r}}|}{|\vec{r}|}\vec{r}. \tag{F.13}$$

Using Eqs. (F.4) and (F.13) yields:

$$\vec{v} = \frac{\dot{a}}{a}\vec{r} = H\vec{r}$$

where H is called the Hubble constant. Since H is positive, this result indicates that the universe is expanding. Although H is constant in space, it is not constant in time where its evolution is described with Eq. (F.6):

$$H(t)^2 = \frac{8\pi G\rho}{3} - \frac{kc^2}{a^2} \tag{F.14}$$

F.4.2   Evolution of the universe

The evolution of the universe can be described by different cosmological models, which involve the combined solution of the Friedmann equation, fluid equation and the equation of state.

Using the fluid equation in Eq. (F.11), where the equation of state for *matter* can be approximated with the pressure $p = 0$, yields:

$$\dot{\rho} + 3\frac{\dot{a}}{a}\rho = 0 \Rightarrow \frac{1}{a^3}\frac{d}{dt}\left(\rho a^3\right) = 0 \Rightarrow \rho = \frac{\rho_0}{a^3} \tag{F.15}$$

Here ρ_0 is a proportionality constant equal to the density at the present time t_0 today. This result simply implies that the density of the universe falls off in proportion to the volume of the universe. Assuming a flat universe with $k = 0$, Eqs. (F.6) and (F.15) further give

$$\dot{a}^2 = \frac{8\pi G \rho_0}{3}\frac{1}{a} \Rightarrow a(t) = \left(\frac{t}{t_0}\right)^{2/3} \quad \text{and} \quad \rho(t) = \frac{\rho_0}{a^3} = \frac{\rho_0 t_0^2}{t^2} \tag{F.16}$$

Hence, $H = \dot{a}/a = 2/(3t)$ and therefore the universe expands forever becoming increasingly slow but never re-collapsing. This derivation is a classic cosmological solution.

On the other hand, if the universe is dominated by *radiation*, $p = \rho c^2/3$ and the fluid equation in Eq. (F.11) becomes:

$$\dot{\rho} + 4\frac{\dot{a}}{a}\rho = 0 \Rightarrow \rho = \frac{\rho_0}{a^4} \tag{F.17}$$

Similarly

$$a(t) = \left(\frac{t}{t_0}\right)^{1/2} \quad \text{and} \quad \rho(t) = \frac{\rho_0}{a^4} = \frac{\rho_0 t_0^2}{t^2} \tag{F.18}$$

Thus, in the case of radiation, the universe expands more slowly although in each case the density falls off as t^2.

The density will be a mixture of both matter and radiation. However, one can consider the case when either radiation or matter dominate using the previous results:

(i) **Radiation dominates:** In this case: $a(t) \propto t^{1/2}$; $\rho_{rad} \propto 1/t^2$; $\rho_{mat} \propto 1/a^3 \propto 1/t^{3/2}$.
(ii) **Matter dominates:** In this case: $a(t) \propto t^{3/2}$; $\rho_{mat} \propto 1/t^2$; $\rho_{rad} \propto 1/a^4 \propto 1/t^{8/3}$.

For the evolution of the universe containing both radiation and matter, matter becomes stable as it becomes dominate over radiation as time progresses. The density of matter will fall off more steeply than that of radiation. As matter starts to dominate, the expansion rate speeds up from a previous $a(t) \propto t^{1/2}$ law to $a(t) \propto t^{2/3}$. This situation is likely the one that applied to our universe until fairly recently.

There are three possible outcomes for the universe as predicted by the Friedmann equation in Eq. (F.6), where $H^2 = (\dot{a}/a)^2$ in accordance with different signs of k. If $k = 0$, as discussed before, the universe will expand forever as $a(t) \propto t^{2/3}$ in Eq. (F.16) but it will slow down at later times as previously mentioned. However, expansion of the universe does not occur when $H = \dot{a}/a = 0$. If $k < 0$, then Eq. (F.6) shows that both terms on the right hand side are positive so that the universe expands forever. On the other hand, when $k > 0$, H becomes zero when the two terms cancel. The kc^2/a^2 term will become more negative relative to the term containing ρ_{mat}. Eventually, expansion will end and a re-collapse of the universe occurs as a result of gravitational attraction. Again, these three possibilities are shown in Fig. F.1.

F.4.3 *Astronomical observations*

For a given value of the Hubble constant H used in Eq. (F.6), a critical density arises for a flat universe when $k = 0$:

$$\rho_c(t) = \frac{3H^2}{8\pi G} \tag{F.19}$$

Equivalently, a dimensionless parameter called the "density parameter", Ω, is defined as

$$\Omega(t) = \frac{\rho}{\rho_c} \tag{F.20}$$

Einstein originally thought that the universe was static. He introduced a cosmological constant that is now used in modern cosmology to better explain the expansion of the universe. It is applied in the Friedmann equation of Eq. (F.6) as an extra term:

$$H^2 = \left(\frac{\dot{a}}{a}\right)^2 = \frac{8\pi G\rho}{3} - \frac{kc^2}{a^2} + \frac{\Lambda}{3} \tag{F.21}$$

By differentiating Eq. (F.21) with respect to time, substituting in $\dot{\rho}$ from Eq. (F.11), and using Eq. (F.21) again, gives the acceleration equation:

$$\frac{\ddot{a}}{a} = -\frac{4\pi G}{3}\left(\rho + \frac{3p}{c^2}\right) + \frac{\Lambda}{3} \tag{F.22}$$

Hence, a positive cosmological constant yields a positive contribution to $\ddot{a}$. This result acts as a repulsive force which can overcome gravitational attraction. This added term can therefore explain the observed acceleration of the universe. Observations further suggest a low-density universe that is nearly flat with a positive cosmological constant. Under these circumstance, the age of the universe from the solution of $a(t)$ is best predicted with $\Omega_0 \simeq 0.3$ and $h \simeq 0.7$. Here Ω_0 is the current density parameter.

It is suggested the matter density is dominated by "dark matter" but that the dark matter density falls short of the critical density. Here the density parameter for the cosmological constant is $\Omega_\Lambda = \Lambda/(3H^2)$. For a flat universe, the condition is $\Omega + \Omega_\Lambda = 1$ with $k = 0$. Thus, favoured modern values for the density parameters are $\Omega \simeq 0.3$ and $\Omega_\Lambda \simeq 0.7$.

F.4.4   *The inflationary universe*

The hot big bang theory however has several difficulties:

(i) **The flatness problem:** The universe has a total density of $\Omega_{\text{tot}} = \Omega_0 + \Omega_\Lambda$, which is close to the critical density where Ω_{tot} is known to be in the range, $0.5 \leq \Omega_{\text{tot}} \leq 1.5$. This suggests that the universe has a geometry close to flat. However, a flat geometry is inherently unstable since if there is any deviation, the universe will become more curved with time. Consequently, for the universe to be close to flat at present, it must have been precisely close to a flat geometry in the beginning.

(ii) **The horizon problem:** Since the universe is of a finite age, and the speed of light is finite, light cannot have travelled from opposite sides of the sky to reach us. On the other hand, all parts of the sky have a microwave background with the same temperature of 2.725 K. This observation indicates thermal equilibrium, requiring some form of communication between different regions of the universe.

(iii) **Particle abundances:** The universe has remained dominated by radiation until at least 1000 years, despite the fact that the radiation density falls off as $1/a^4$ much faster than that of matter. If the universe started off with any small type of matter, it would have rapidly reached prominence. For instance, while magnetic monopoles (and other type of exotic particles) are relic particles important at the early time of the big bang, the universe is not dominated by these particles now.

The solution to these three problems was proposed in 1981 by Alan Guth. Here inflation occurred in the early evolution of the universe as shown in Fig. F.3, with an accelerating scale factor, i.e., $\ddot{a} > 0$.

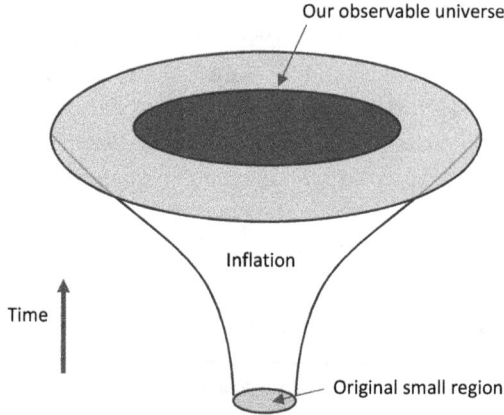

Figure F.3. Schematic of the inflationary model that solves the flatness, horizon and particle abundance problems of the hot big bang model.

Hence, from the Friedmann equation in Eq. (F.21), the first two terms are rapidly reduced by expansion while the last one remains constant so that:

$$H^2 = \frac{\Lambda}{3} \Rightarrow \dot{a} = \sqrt{\frac{\Lambda}{3}} a \Rightarrow a(t) = \exp\left(\sqrt{\frac{\Lambda}{3}} t\right) \qquad \text{(F.23)}$$

While the universe is dominated by a cosmological constant, a dramatic expansion rate can occur at 10^{-34} s with an expansion factor of 10^{27} or more. Inflation stops after the energy in the cosmological constant is changed into conventional matter. During cooling, the universe could have undergone a phase transition with matter taking the form of a scalar field. Once the phase transition ends, the scalar field decays away and the inflationary expansion ceases.

F.4.5 *The cosmological constant problem*

The "cosmological constant problem" arises when comparing results from General Relativity and Quantum Field Theory in regards to the gravitational properties of the vacuum [Wang *et al.*, 2017]. Einstein's field equations for the vacuum follow from Eq. (B.39b), where natural units $c = \hbar = 1$ as well as the metric signature set $(-+++)$ are assumed in the following analysis:

$$G_{\mu\nu} + \Lambda g_{\mu\nu} = 8\pi G T_{\mu\nu} \qquad \text{(F.24)}$$

The vacuum stress-energy tensor is proportional to $\eta_{\mu\nu}$ for Minkowski space–time:

$$T_{\mu\nu}^{\text{vac}}(t, \boldsymbol{x}) = -\rho^{\text{vac}}\eta_{\mu\nu} \tag{F.25}$$

This relation can be generalized to curved space–time as

$$T_{\mu\nu}^{\text{vac}}(t, \boldsymbol{x}) = -\rho^{\text{vac}}g_{\mu\nu}(t, \boldsymbol{x}) \tag{F.26}$$

For conservation of the stress-energy tensor $\nabla^{\mu}T_{\mu\nu}^{\text{vac}} = 0$, the density must be constant. The form of Eq. (F.26) is in fact equivalent to that of the cosmological constant. In particular, on moving the term $\Lambda g_{\mu\nu}$ in Eq. (F.24) to the right-hand side yields

$$G_{\mu\nu} = -8\pi G \rho_{\text{eff}}^{\text{vac}} g_{\mu\nu} \tag{F.27}$$

where

$$\rho_{\text{eff}}^{\text{vac}} = \rho^{\text{vac}} + \frac{\Lambda}{8\pi G} \tag{F.28}$$

Thus, the cosmological constant acts like a source of vacuum energy and contributes to the total effective vacuum energy density. The total effective vacuum energy density $\rho_{\text{eff}}^{\text{vac}}$ can be obtained by large scale cosmological observations in terms of "dark energy" that describes the acceleration of the universe.

The Friedmann–Lemâitre–Robertson–Walker (FLRW) metric is described in Appendix F.3 for a homogeneous and isotropic universe. It can be assumed in the following analysis that the universe is spatially flat (such that $k = 0$). Given the Hubble expansion rate $H = \dot{a}/a$, the Friedmann equation in Eq. (F.21) can be written as

$$3H^2 = 8\pi G \rho_{\text{eff}}^{\text{vac}} \equiv \Lambda_{\text{eff}} \tag{F.29}$$

Similarly, Eq. (F.22) for the acceleration scale factor can be written as:

$$\ddot{a} = \frac{8\pi G \rho_{\text{eff}}^{\text{vac}}}{3}a = \frac{\Lambda_{\text{eff}}}{3}a \tag{F.30}$$

In the derivation of Eq. (F.30), the cosmological constant can be interpreted as arising from a form of energy which has a negative pressure equal in magnitude to its (positive) energy density (i.e., $p = -\rho^{\text{vac}}$). The solution of Eq. (F.30) is

$$a(t) = a(0)e^{Ht} \tag{F.31}$$

Here H is the defined by the initial value constant in Eq. (F.29).

In accordance with the Lambda-Cold Dark Matter (CDM) model for big bang cosmology in Chapter 6, the effective cosmological constant is responsible for the accelerated expansion of the universe with Eq. (F.30). It contributes about 69% of the current Hubble expansion rate where from Eq. (F.29):

$$\Lambda_{\text{eff}} = 3\Omega_\Lambda H_0^2 \approx 4.32 \times 10^{-84} \ (\text{GeV})^2 \tag{F.32}$$

where $\Omega_\Lambda = 0.69$ is the dark energy density parameter and H_o is the current Hubble constant. Equivalently, from Eq. (F.20),

$$\rho_{\text{eff}}^{\text{vac}} = \Omega_\Lambda \rho_c \approx 2.57 \times 10^{-47} \ (\text{GeV})^4 \tag{F.33}$$

where the critical density is defined in Eq. (F.19) using the current Hubble constant.

In comparison, the predicted energy density from quantum field theory is significantly larger [Wang *et al.*, 2017]. Contributions to the energy density arise from various sources including: the zero point energies of all quantum fields due to vacuum fluctuations (i.e., $\sim 10^{72} \ (\text{GeV})^4$), phase transitions from spontaneous symmetry breaking in the electroweak theory (i.e., $\sim 10^9 \ (\text{GeV})^4$) as well as other phase transitions in the early universe (i.e., chiral symmetry breaking in quantum chromodynamics theory ($\sim 10^{-2} \ (\text{GeV})^4$) and grand unification ($\sim 10^{64} \ (\text{GeV})^4$)). These contributions are larger than the observed value in Eq. (F.33) by some 50 to 120 orders of magnitude. Hence, the cosmological constant problem lies in the fact that one has to take the cosmological constant Λ to a precision of 50 decimal places to cancel the excess vacuum energy density.

For example, [Wang *et al.*, 2017] have addressed the cosmological constant problem by considering a vacuum energy density as precisely predicted by quantum field theory without renormalization. It is suggested that this quantity gravitates in accordance with the equivalence principle of general relativity. However, the vacuum fluctuations have a constantly fluctuating and an extremely inhomogeneous vacuum energy density. The quantum vacuum gravitates differently from a cosmological constant. However, in accordance with the concept of a space–time foam at the Planck scale as originally suggested by Wheeler (i.e., a structure similar to that considered in loop quantum gravity theory in Appendix C.3), each spatial point in space–time is considered to be sourced by a vacuum that oscillates alternatively between expansion and contraction. In this situation, the phases of neighbouring points are different. Although the gravitational effect produced by the vacuum energy is still huge at the sufficiently small Planck scale, its effect at the macroscopic level is largely cancelled out. However, due to a weak parametric resonance, the expansion outweighs the contraction slightly during each oscillation. As such, this effect accumulates on a large cosmological scale resulting in the observable expansion of the universe.

G   Black Holes

The mechanics of black holes and their correspondence to thermodynamics theory are described in Appendix G.1. The "Schwarzschild metric" is presented in Appendix G.2 as a spherically-symmetric solution of the vacuum field equations of general relativity. This solution can be applied outside a spherical mass for a static black hole that has neither electric charge nor angular momentum. This metric reveals a coordinate singularity at the event horizon of a black hole. The properties of the black hole are derived including its temperature and entropy. Finally, the characteristics of the event horizon is described mathematically in Appendix G.3 for infalling and outgoing particles and light. In addition, coordinate transformations are used for determination of the physical singularity at the centre of the black hole.

G.1   Black Hole Mechanics

The laws of black hole mechanics align with those of thermodynamics as shown in the early 1970s by James Bardeen, Brandon Carter and Stephen Hawking. The zeroth law of black hole mechanics recognizes that the surface gravity κ at the horizon of a stationary black hole is constant. The first law relates the energy E, the horizon area A, the angular momentum J and electric charge Q of a black hole:

$$dE = \frac{\kappa}{8\pi}dA + \Omega dJ + \Phi dQ$$

where Ω is the angular velocity and Φ is the electrostatic potential. The charge and angular momentum of a black hole are constrained by its mass m such that:

$$\frac{Q^2}{4\pi\epsilon_0} + \frac{c^2 J^2}{Gm^2} \leq Gm^2$$

where G is the universal gravitational constant and ϵ_0 is the vacuum permittivity. Assuming a weak energy condition, the second law states that the horizon area is an increasing function of time such that $\frac{dA}{dt} \geq 0$. However, this law was superseded by the occurrence of Hawking radiation that causes the mass and area of its horizon to decrease over time. The second law is analogous to the second law of thermodynamics where the entropy of a closed system is a non-decreasing function of time. Hence, if black holes with horizon areas A_1 and A_2 coalesce, the new black hole with area A_3 must satisfy the relation $A_3 > A_1 + A_2$. The third law indicates that a black hole cannot form with vanishing surface gravity (i.e., $\kappa = 0$).

This correspondence with thermodynamics relates the area of the horizon A with the entropy S of the black hole and the surface gravity κ with its temperature T. The entropy of a black hole can either be expressed in terms of its mass or area. In particular, the entropy of a black hole is proportional to the square of its mass. The entropy is also related to the horizon area in units of the Planck length ℓ_p. Here the Planck length is defined as $\ell_p = \sqrt{\frac{G\hbar}{c^3}} \sim 10^{-35}$ m, where $\hbar$ is the reduced Plank's constant. Thus, this correspondence gives:

$$S = \frac{A}{4\ell_p^2}$$

G.2   Field Equations

The set of field equations from the general theory of relativity that relate the curvature of space–time to the matter-energy content of the universe was given in Eq. (B.39a) and is reproduced below:

$$R_{\mu\nu} - \frac{1}{2}Rg_{\mu\nu} + \Lambda g_{\mu\nu} = \frac{8\pi G}{c^4}T_{\mu\nu} \qquad (\text{G.1})$$

where

- $R_{\mu\nu}$ is the Ricci tensor that accounts for the curvature of space as derived from the Riemann curvature tensor $R_{\alpha\beta} = R^{\mu}_{\alpha\mu\beta}$;
- $g_{\mu\nu}$ is the metric tensor that describes the geometry of space–time;
- R is the Ricci scalar computed from contraction of the Ricci tensor;
- Λ is the cosmological constant;
- G is the universal constant for gravitation;
- $T_{\mu\nu}$ is the energy-momentum tensor.

To study the gravitational field outside of a source, the energy-momentum tensor $T_{\mu\nu} = 0$ and the cosmological constant Λ can be set to zero in the situation that there is no matter or energy present in the space–time. Hence, the equations become:

$$R_{\mu\nu} - \frac{1}{2}Rg_{\mu\nu} = 0 \qquad (\text{G.2})$$

These vacuum field equations can be used to study the space–time outside of a massive body such as a black hole. This type of study is possible because all of the mass is concentrated in a single point in the singularity at the centre of the black hole.

Consider a simple case of a non-rotating black hole of mass m that is spherically symmetric. As shown in Eq. (B.66), the Schwarzschild metric can be derived from a solution of the vacuum field equations which describe the space–time outside of a black hole:

$$ds^2 = \left(1 - \frac{r_s}{r}\right) c^2 dt^2 - \frac{dr^2}{\left(1 - \frac{r_s}{r}\right)} - r^2 \left(d\theta^2 + \sin^2\theta d\phi^2\right) \qquad (\text{G.3})$$

The so-called Schwarzschild radius is given by $r_s = 2Gm/c^2$, which depicts the location of the event horizon of the black hole. This point appears to be a singularity because the coefficient of dr^2 goes to infinity when $r = r_s$. However, this singularity is actually an artifact of the coordinate system. In particular, a calculation of the invariant quantity

$$R^{\mu\nu\rho\sigma} R_{\mu\nu\rho\sigma} = \frac{12r_s^2}{r^6}$$

reveals that only $r = 0$ is a true physical singularity.

Although r_s is not a singularity, it is an important location that denotes the event horizon of the black hole. This boundary consequentially divides space–time into an external world and the point of no return. Nothing that crosses the event horizon can return to the outside universe. Thus, black holes are black because not even light can escape from inside the event horizon.

Black Hole Properties: The temperature and entropy of a static non-charged Schwarzschild black hole is evaluated in the following sections.

(i) **Temperature:** Consider a "Wick's rotation" where $t \to i\tau$ for the Schwarzschild metric in Eq. (G.3), which gives:

$$ds^2 = -\left(1 - \frac{r_s}{r}\right) c^2 d\tau^2 - \frac{dr^2}{\left(1 - \frac{r_s}{r}\right)} - r^2 \left(d\theta^2 + \sin^2\theta d\phi^2\right) \qquad (\text{G.4})$$

The following variable changes can be applied that still capture the line metric quantities for $d\tau$ and dr [McMahon, 2009]:

$$R'd\alpha = \left(1 - \frac{r_s}{r}\right)^{1/2} c d\tau$$

and

$$dR' = \left(1 - \frac{r_s}{r}\right)^{-1/2} dr$$

For $r > r_s$, considering a calculation of first order near the event horizon where $r \sim r_s$, the first relation can be rewritten and simplified as

$$R'd\alpha = \frac{1}{r^{1/2}} \left(r - r_s\right)^{1/2} c d\tau \sim \frac{1}{r_s^{1/2}} \left(r - r_s\right)^{1/2} c d\tau$$

Integrating this latter equation on both sides with the limits for $\alpha \to 0$ to 2π, and for $\tau \to 0$ to β, yields

$$2\pi R' = r_s^{-1/2} \left(r - r_s\right)^{1/2} c\beta \qquad (G.5)$$

Similarly, the second relation can be written as

$$dR' = r^{1/2} \frac{dr}{\left(r - r_s\right)^{1/2}} \sim r_s^{1/2} \frac{dr}{\left(r - r_s\right)^{1/2}}$$

with the limits for $r \to r_s$ to r. Letting $u = (r - r_s)$, $du = dr$ and the limits for u become: $u \to 0$ to u. Thus, integrating:

$$R' = r_s^{1/2} \int_0^u u^{-1/2} du = 2r_s^{1/2} u^{1/2} = 2r_s^{1/2} \left(r - r_s\right)^{1/2} \qquad (G.6)$$

Dividing Eq. (G.5) by Eq. (G.6) gives

$$2\pi = \frac{c\beta}{2r_s} \Rightarrow c\beta = 4\pi r_s = \frac{8\pi Gm}{c^2}$$

Recognizing that β is the same parameter from thermodynamics such that $\beta = 1/(k_B T)$ where T is the temperature and k_B is the Boltzmann's constant, the temperature for the black hole is derived as

$$T = \frac{c^3}{8\pi G m k_B}$$

For a temperature in the physical units of Kelvin, the formula can be multiplied by a simple conversion factor. Using the reduced Planck's constant for this conversion factor, $\hbar$, yields the exact expression as derived by Hawking for the temperature of a black hole:

$$T = \frac{\hbar c^3}{8\pi G m k_B}$$

(G.7)

(ii) **Entropy:** Using Eq. (G.7) for the temperature, the entropy of the black hole can be further determined. From the second law of thermodynamics, $dE = T dS$, where using the energy $E = mc^2$ for a static uncharged black hole:

$$c^2 dm = T dS \Rightarrow c^2 \int m \, dm = \frac{\hbar c^3}{8\pi G k_B} \int dS$$

Integrating the above expression yields: $\frac{c^2 m^2}{2} = \frac{\hbar c^3}{8\pi G k_B} S$. Hence, solving for S:

$$S = \frac{4\pi m^2 G k_B}{\hbar c}$$

(G.8)

Thus, the entropy is proportional to the square of the mass of the black hole. Moreover, in dimensionless units, $S = \frac{4\pi m^2 G}{\hbar c} = \frac{A}{4\ell_p^2}$, where, by definition, the Planck length $\ell_p = \sqrt{\frac{G\hbar}{c^3}}$, the area of the event horizon $A = 4\pi r_s^2$ and the Schwarzschild radius $r_s = \frac{2Gm}{c^2}$. Thus, the entropy of the black hole is equivalently proportional to the area of the event horizon.

G.3   Event Horizon

For the Schwarzschild metric in Eq. (G.3), when $r > r_s$, the line element for the time component is positive while the radial line element is negative. However, for $r < r_s$ the signs of the metric reverse. Hence, a world-line along the t-axis has $ds^2 < 0$ and therefore follows a space-like curve. On the other hand, along the r-axis, $ds^2 > 0$ so that a time-like curve is followed. Since the spatial and temporal character of the coordinates has reversed, a particle inside the Schwarzschild radius cannot remain stationary at constant distance r.

Consider an infalling particle from infinity travelling toward a black hole with a Schwarzschild geometry. This particle will move on the path

[McMahon, 2006]:

$$\left(1 - \frac{r_s}{r}\right)\frac{dt}{d\tau} = 1 \text{ and } \left(\frac{dr}{d\tau}\right)^2 = \frac{r_s}{r}$$

Hence:

$$\frac{dr/d\tau}{dt/d\tau} = \frac{dr}{dt} = -\sqrt{\frac{r_s}{r}}\left(1 - \frac{r_s}{r}\right)$$

where a minus sign is taken for an infalling body. Letting $x = \frac{r_s}{r}$ implies $\frac{dx}{dr} = -\frac{r_s}{r^2}$ and $dr = -\frac{r_s}{x^2}dx$. Using these relations:

$$\frac{dr}{dt} = -\sqrt{\frac{r_s}{r}}\left(1 - \frac{r_s}{r}\right) \Rightarrow dt = \frac{r_s dx}{x^{5/2}(1 - x)}$$

One can integrate over the limits t_0 to t and x_0 to x such that:

$$\int_{t_0}^{t} dt = r_s \int_{x_0}^{x} \frac{dx}{x^{5/2}(1 - x)}$$

Using Maple [Maplesoft, 2015]:

$$\int \frac{dx}{x^{5/2}(1 - x)} = -\frac{2}{3x^{3/2}} + \ln\frac{(\sqrt{x} + 1)}{(\sqrt{x} - 1)} - \frac{2}{\sqrt{x}}$$

and the following expression is obtained after factoring:

$$t - t_0 = \frac{2}{3\sqrt{r_s}}\left[r_0^{3/2} - r^{3/2} + 3r_s\left(\sqrt{r_0} - \sqrt{r}\right)\right]$$

$$- r_s \ln\left[\frac{(\sqrt{r_0} + \sqrt{r_s})(\sqrt{r} - \sqrt{r_s})}{(\sqrt{r} + \sqrt{r_s})(\sqrt{r_0} - \sqrt{r_s})}\right]$$

The term $\sqrt{r} - \sqrt{r_s} \to 0$ as $r \to r_s$ in the numerator of the natural logarithm function. Given that $\ln(0) = -\infty$, the time $t \to \infty$ as the event horizon is approached. Thus, the event horizon is never passed. For an outside observer far from the black hole, a falling body will never reach the event horizon at $r = r_s$. On the other hand, using the proper time for the radially infalling body: $\frac{dr}{d\tau} = -\sqrt{\frac{r_s}{r}}$, and assuming that this body starts at position $r = r_0$ at proper time $\tau = \tau_0$, the resulting differential equation is separable and can be integrated as:

$$\frac{1}{\sqrt{r_s}}\int_{r}^{r_0}\sqrt{r}\,dr = \int_{\tau_0}^{\tau} d\tau \Rightarrow \tau - \tau_0 = \frac{2}{3\sqrt{r_s}}\left(r_0^{3/2} - r^{3/2}\right)$$

Hence, in this result, there is no appearance of the surface $r = r_s$. A body will fall continuously to $r = 0$ in a finite proper time, in contrast to that seen by the distant observer.

The nature of the singularity in space–time at $r = r_s$ and inside the event horizon at $r < r_s$ requires a coordinate transformation to remove the coordinate singularity. Hence, this behaviour can be investigated further using (i) *Eddington–Finkelstein* and (ii) *Kruskal–Szekeres* coordinate transformations.

(i) **Eddington–Finkelstein coordinates:** Consider only radial paths in the Schwarzschild metric so that $d\theta = d\phi = 0$ and therefore:

$$ds^2 = \left(1 - \frac{r_s}{r}\right) dt^2 - \frac{dr^2}{\left(1 - \frac{r_s}{r}\right)} \tag{G.9}$$

For a light ray, $ds^2 = 0$. Thus, the slope of a light cone is

$$\frac{dt}{dr} = \pm \left(1 - \frac{r_s}{r}\right)^{-1} \tag{G.10}$$

Far away from $r = r_s$, as $r \to \infty$, $\frac{dt}{dr} = \pm 1$. Thus, one recovers the motion of light rays in flat space where integration gives $t = \pm r$ (up to a constant). Next, consider an outgoing light ray (positive sign) near $r = r_s$. Thus, $\frac{dt}{dr} = \frac{r}{r - r_s}$. When $r \to r_s$, $dt/dr \to \infty$. As such, light cones become more narrow. The cone eventually becomes a vertical line at $r = r_s$ as shown in Fig. G.1.

The singularity at $r = r_s$ can in fact be removed on integrating Eq. (G.10) to obtain a function $t(r)$. For outgoing radial null curves (up to a constant):

$$t = r + r_s \ln |r - r_s|$$

This result suggests a "tortoise" coordinate transformation so that the metric may be written in a form that only shows a singularity at the origin. A new coordinate r^* can therefore be defined:

$$r* = r + r_s \ln \left(\frac{r}{r_s} - 1\right) \tag{G.11}$$

along with the null coordinates

$$u = t - r^* \quad \text{and} \quad \nu = t + r^* \tag{G.12}$$

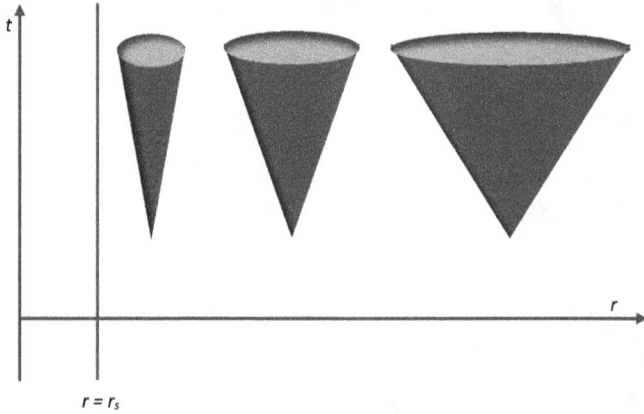

Figure G.1. In Schwarzschild coordinates, light cones become thinner as they approach the event horizon.

From Eq. (G.11), $dr^* = \frac{dr}{1-r_s/r}$. From Eq. (G.12), $dt = d\nu - dr^* = d\nu - \frac{dr}{1-r_s/r} \Rightarrow dt^2 = d\nu^2 - \frac{2d\nu dr}{(1-r_s/r)} + \frac{dr^2}{(1-r_s/r)^2}$. Substituting these expressions into the Schwarzschild metric of Eq. (G.3) gives:

$$ds^2 = \left(1 - \frac{r_s}{r}\right) d\nu^2 - 2d\nu dr - r^2 \left(d\theta^2 + \sin^2 \theta d\phi^2\right) \qquad (G.13)$$

Although the singularity at $r = 0$ still occurs, the coordinate singularity at $r = r_s$ has been removed.

Again, consider a radial light path for Eq. (G.13) where $d\theta = d\phi = ds^2 = 0$. Thus, dividing this resultant equation through by $d\nu^2$ gives

$$\left(1 - \frac{r_s}{r}\right) - 2\frac{dr}{d\nu} = 0 \qquad (G.14)$$

If $r = r_s$ then $dr/d\nu = 0 \Rightarrow r(\nu) = $ constant. This condition describes stationary light rays that are neither outgoing nor ingoing. Rearranging Eq. (G.14) gives

$$\frac{d\nu}{dr} = \frac{2}{\left(1 - \frac{r_s}{r}\right)}$$

Integrating this equation yields:

$$\nu(r) = 2(r + r_s \ln |r - r_s|) + \text{constant}$$

This equation provides the paths that radial light rays travel on in a (ν, r)-coordinate system. If $r > r_s$, then ν increases as r increases,

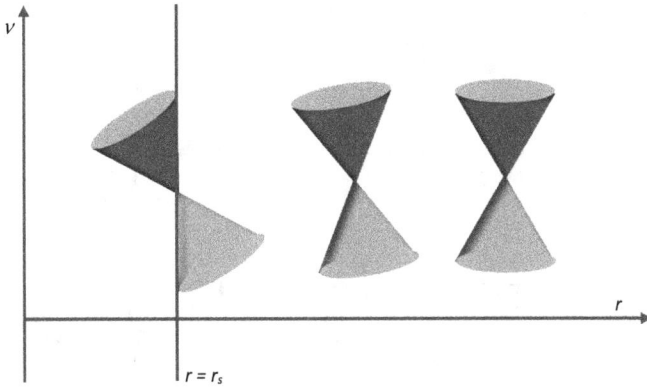

Figure G.2. In Eddington–Finkelstein coordinates, the coordinate singularity at $r = r_s$ is removed. As r becomes smaller, light cones tip over. For $r = r_s$ all geodesics that are directed toward the future head toward the singularity at $r = 0$.

which provides the expected behaviour for outgoing light rays. On the other hand, for $r < r_s$, ν increases as r decreases, indicating ingoing light rays.

In this coordinate system, light cones do not become increasingly narrow but instead pass through the location at $r = r_s$. However, importantly, since the time and radial coordinates reverse their character inside the event horizon, the light cones tilt over in this region as shown in Fig. G.2.

In summary:

- There is a coordinate singularity at the location $r = r_s$; however, this singularity can be removed with a change of coordinates.
- The surface at $r = r_s$ is the event horizon that represents a one-way membrane. Future-directed light-like and time-like curves can cross the event horizon but the reverse is not possible.
- Moving in the direction of smaller r, light cones begin to tip over. At $r = r_s$, outward travelling photons remain stationary.
- Inside the event horizon, future-directed light-like and time-like curves are directed toward the central physical singularity at $r = 0$.
- The Schwarzschild coordinate system is useful for describing the geometry outside of the event horizon for all time. Another coordinate system is needed to describe the locations at $r \leq r_s$.

(ii) **Kruskal–Szekeres coordinates:** Kruskal–Szekeres coordinates can be used to extend the Schwarzschild geometry into the region $r < r_s$.

Thus, defining the new coordinates X and T, which serve as global radial and time markers, respectively:

For $r > r_s$:

$$X = e^{r/(2r_s)}\sqrt{\frac{r}{r_s} - 1}\cosh\left(\frac{ct}{2r_s}\right), \quad T = e^{r/(2r_s)}\sqrt{\frac{r}{r_s} - 1}\sinh\left(\frac{ct}{2r_s}\right)$$

$$(G.15)$$

and for $r < r_s$:

$$X = e^{r/(2r_s)}\sqrt{1 - \frac{r}{r_s}}\sinh\left(\frac{ct}{2r_s}\right), \quad T = e^{r/(2r_s)}\sqrt{1 - \frac{r}{r_s}}\cosh\left(\frac{ct}{2r_s}\right)$$

$$(G.16)$$

The Kruskal–Szekeres form of the metric becomes:

$$ds^2 = \left(\frac{4r_s^3}{r}\right)e^{-r/r_s}\left(dX^2 - dT^2\right) + r^2\left(d\theta^2 + \sin^2\theta d\phi^2\right) \quad (G.17)$$

These coordinates are shown in Fig. G.3, where the regions O and O' are outside of the event horizon. The regions I and I' correspond to interior regions $r < r_s$. The hyperbola $r = $ constant is for a constant radius outside of $r = r_s$. There is still a real singularity at $r = 0$ but no coordinate singularity at $r = r_s$. The dashed line indicated by A depicts a light ray travelling inward with a slope of -1 that must hit the singularity at $r = 0$.

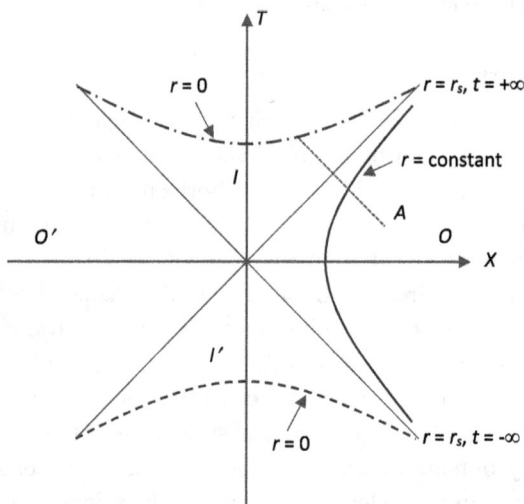

Figure G.3. Graphical depiction of the Kruskal–Szekeres coordinates.

H Glossary

Baryons: A type of subatomic particle that contains an odd number of quarks. These particles belong to the hadron family of particles. Baryons are classified as fermions since they have a half-integral spin.

Baryon Number: A quantum number for elementary particles. Baryons have a number one, antibaryons a number minus one, and all other observable particles have a number zero. Quarks have a baryon number of $1/3$ and antiquarks a number of $-1/3$.

Big Bang: The theory that the expanding universe began at a finite time in the past in an initial state of high pressure and density.

Black Body Radiation: Electromagnetic radiation in thermal equilibrium with the same energy density in each wavelength range as the radiation emitted from a totally absorbing heated body.

Black Hole: A black hole is a region of space–time from which gravity is so strong that no particle nor electromagnetic radiation can escape.

Boson: Subatomic particles whose spin is an integer value.

Colour Force: The strong force, which is one of the four fundamental forces in nature that holds quarks together.

Comoving Coordinates: A coordinate system which expands in the same way as space.

Cosmology: A scientific study of the origin of the universe, its large scale structures and dynamics, and the ultimate fate of the universe.

Cosmological Constant: A constant coefficient that Einstein temporarily added to his field equations of general relativity. It has been reinterpreted as the energy density of space.

Cosmological Principle: The hypothesis that the universe is homogeneous and isotropic.

Dark Energy: Unknown form of energy that causes the accelerating expansion of the universe.

Dark Matter: A hypothetical form of matter believed to account for 85% of the matter in the universe that does not interact with electromagnetic radiation. It has been postulated to explain some astrophysical observations.

Electromagnetic Force: One of the four fundamental forces in nature, carried by electric and magnetic fields, that accounts for the physical interaction between electrically charged particles.

Electron: The lightest subatomic particle with mass that carries a negative single electric charge. Electrons belong to the first generation of lepton particles.

Electroweak Interaction: The unified description of electromagnetism and the weak interaction.

Event Horizon: The surface of a black hole that defines a boundary where the velocity needed to escape the black hole exceeds that of light.

Fermion: A particle that has a half-odd-integer spin that obeys the Pauli exclusion principle.

Feynman Diagram: Diagram which symbolizes mathematical expressions that describe the behaviour and interaction of subatomic particles.

Friedmann Model: Mathematical model that governs the expansion of the universe within the context of general relativity based on the Cosmological Principle.

Gauge Theory: A class of field theories describing weak, electromagnetic and strong interactions that are invariant under symmetry transformations whose effect varies point by point in space–time.

General Relativity: A theory of gravity developed by Albert Einstein which embodies the concept that gravity is a curvature of the space–time continuum.

Gluon: An exchange particle (or gauge boson) that binds quarks together as the strong force forming hadrons such as protons and neutrons.

Gravitational Force: One of the four fundamental forces of nature that causes the mutual attraction of all things with mass and energy. It determines the motion of planets, stars, galaxies and even light.

Gravitational Waves: Disturbances in the curvature of space–time that propagate outwards from their source as waves at the speed of light. These waves are analogous to light waves in the electromagnetic field.

Hadron: A subatomic particle that participates in the strong interaction. Hadrons are divided into baryons that are made up of an an odd number of quarks and mesons that are made up of an even number (usually one quark and one anti-quark). "Exotic" hadrons containing more than three quarks have also been discovered.

Hawking Radiation: Thermal radiation that is released from the event horizon of a black hole.

Hermitian Operator: An operator, representing physical variables in quantum mechanics such as position, momentum and energy, which has real eigenvalues that can be related to experimentally-determined observables. A quantum mechanical operator that is equal to its own adjoint.

Higgs Boson: An elementary massive particle in the Standard Model of particle physics produced by a quantum excitation of the Higgs field. It is a scalar boson with zero spin, positive parity, no electric or colour charge, and interacts with mass.

Hubble's Law: Observation that distant galaxies are moving away from the Earth at speeds proportional to their distance from the Earth.

Lagrangian: A quantity that characterizes the state of a physical system. It is defined as the kinetic energy minus the potential energy.

Lepton: An elementary particle of half-integer spin that does not undergo strong interactions.

Light Cone: The path that a ray of light takes emanating from a single event, which is a localized point in space and time and travels in all directions.

Loop Quantum Gravity: A theory of quantum gravity that combines quantum mechanics with the general theory of relativity.

M Theory: General theory that provides for a unification of all consistent versions of string theory.

Meson: A class of strongly interacting particles, including pi mesons, K-mesons, rho mesons among others with zero baryon number.

Metric Tensor (Metric): Fundamental mathematical quantity used in the theory of general relativity that captures the geometric and casual structure of space–time.

Muons: An unstable elementary particle of negative charge similar to the electron but 207 times more massive.

Neutrino: A subatomic particle that is similar to an electron but has no electrical charge with a very small mass having only weak and gravitational interactions.

Neutron: An uncharged particle found along with protons in ordinary atomic nuclei.

Neutron Star: The collapsed core of a massive supergiant star composed almost entirely of neutrons.

Pauli Exclusion Principle: This principle requires that two or more identical particles with half-integer spins (i.e., fermion particles such as electrons) cannot occupy the same quantum state in a given system.

Photon: A particle associated with light.

Pion: A hadron with the lowest mass. There are three types of pions, a positively-charged particle, an electrically-charged particle and a neutral one. Also known as pi mesons.

Principle of Least Action: The action represents an integral over time of a Lagrangian function for evaluation of the development of a system from an initial to a final time. In the principle of least action, this quantity is minimized (or maximized) in order to find an optimum quantity of interest.

Proper Time: The time measured by a clock following a world-line that someone carries in their own rest frame.

Proton: A positively charged particle found along with neutrons in ordinary atomic nuclei.

Pulsar: A highly-magnetized rotating neutron star, which emits beams of electromagnetic radiation out of its magnetic poles.

Quantum Chromodynamics: The theory of strong interactions between quarks as mediated by gluons.

Quantum Electrodynamics: The relativistic quantum field theory of electrodynamics.

Quantum Field Theory: A theoretical framework combining classical field theory, special relativity and quantum mechanics. It is used to construct models of subatomic particles in particle physics and for condensed matter physics. In this theory, particles are treated as excited states (or quanta) of the underlying quantum fields.

Quantum Mechanics: A physical theory that replaces classical physics where energy, momentum, angular momentum among other quantities of a bound system are restricted to discrete values. Objects have both wave-like and particle-like behaviour. There is a limit as to how accurately physical quantities can be predicted prior to their measurement.

Quark: An elementary particle that is a fundamental constituent of matter. Quarks can combine to form composite hadrons, the most stable of which are protons and neutrons.

Quasar: An astronomical object of very high luminosity found in the centre of galaxies and powered by gas spiralling into a super-massive black hole.

Red-shift: The observed shift of spectral lines of distant astronomical bodies towards longer wavelengths. A gravitational red-shift is a phenomenon where electromagnetic waves and photons lose energy as they travel out of a gravitational well.

Renormalization: A procedure in quantum field theory in which divergent calculations containing infinities can be eliminated with a redefinition using measurable quantities.

Robertson-Walker Metric: A metric based on the exact solution of Einstein's field equations of general relativity.

Schwarzschild Radius: The radius of the event horizon of a static black hole for Schwarzschild's solution of Einstein's field equations of general relativity.

Special Relativity: Albert Einstein's theory about the relationship between space and time where the laws of physics are identical (invariant) in all (inertial) frames of reference with no acceleration. The speed of light in vacuum is a constant for all observers.

Spontaneous Symmetry Breaking: A spontaneous process where a physical system in a symmetric state ends up in an asymmetric state. This process is important in particle physics for the force carrier particles with gauge symmetry.

Standard Model of Particle Physics: The theory describing three of the four fundamental forces in the universe with a classification of all known elementary particles.

String Theory: Theoretical framework where point-like particles are replaced by one-dimensional strings.

Strong Force: One of the four fundamental forces of nature that confines quarks into protons, neutrons and other hadron particles. The strong interaction binds neutrons and protons together to create atomic nuclei.

Uncertainty Principle: Fundamental limit in quantum mechanics in which the accuracy of physical quantities such as position and momentum can be predicted from initial conditions.

Vacuum Expectation Value: Average or expectation value of an operator in quantum field theory. The Higgs field has a non-zero expectation value, which underlies the Higgs mechanism with spontaneous symmetry breaking in the Standard Model.

W and Z Vector Bosons: These particles (with respective symbols W^+, W^- and Z^0) mediate the weak force.

Weak Force: One of the four fundamental forces of nature that governs the process of radioactivity.

Bibliography

[Adler et al., 1975] Adler, R., Bazin, M., and Schiffer, M. (1975). *Introduction to General Relativity*. International Series in Pure and Applied Science. McGraw Hill, Inc.

[Bjorken and Drell, 1964] Bjorken, J. D. and Drell, S. D. (1964). *Relativistic Quantum Mechanics*. McGraw Hill.

[Dicke and Wittke, 1960] Dicke, R. H. and Wittke, J. P. (1960). *Introduction to Quantum Mechanics*. Addison-Wesley Publishing.

[d'Inverno, 1992] d'Inverno, R. (1992). *Introducing Einstein's Relativity*. Oxford University Press.

[Dodelson, 2003] Dodelson, S. (2003). *Modern Cosmology*. Academic Press.

[Einstein, 2014] Einstein, A. (2014). *The Meaning of Relativity*. New Princeton Science Library Edition. Princeton University Press.

[Eisberg and Resnik, 1974] Eisberg, R. and Resnik, R. (1974). *Quantum Physics of Atoms, Molecules, Solids, Nuclei, and Particles*. John Wiley and Sons, Inc.

[Gamow, 1967] Gamow, G. (1967). *One, Two, Three...Infinity, Facts and Speculations of Science*. Bantam Science and Mathematics Edition. Bantam Books, Inc.

[Goldstein, 1965] Goldstein, H. (1965). *Classical Mechanics*. Addison-Wesley Publishing.

[Gubser, 2010] Gubser, S. S. (2010). *The Little Book of String Theory*. Princeton University Press.

[Hawking, 1988] Hawking, S. W. (1988). *A Brief History of Time, From the Big Bang to Black Holes*. Bantam Books. Bantam Doubleday Dell Publishing Group, Inc.

[Klauber, 2013] Klauber, R. D. (2013). *Student Friendly Quantum Field Theory*. Sandtrove Press.

[Lewis et al., 2017] Lewis, B. J., Onder, E. N., and Prudil, A. A. (2017). *Fundamentals of Nuclear Engineering*. Wiley international editions: Nuclear engineering. Wiley.

[Lewis et al., 2022] Lewis, B. J., Onder, E. N., and Prudil, A. A. (2022). *Advanced Mathematics for Engineering Students, The Essential Toolbox*. Butterworth-Heinemann.

[Liddle, 2015] Liddle, A. (2015). *An Introduction to Modern Cosmology.* John Wiley and Sons, Inc.

[Lieber and Lieber, 1966] Lieber, L. R. and Lieber, H. G. (1966). *The Einstein Theory of Relativity.* Holt, Rinehart and Winston.

[Maplesoft, 2015] Maplesoft (2011–2015). Maple programming guide. http://www.maplesoft.com/.

[McMahon, 2006] McMahon, D. (2006). *Relativity Demystified.* McGraw Hill.

[McMahon, 2008] McMahon, D. (2008). *Quantum Field Theory Demystified.* McGraw Hill.

[McMahon, 2009] McMahon, D. (2009). *String Theory Demystified.* McGraw Hill.

[Meisner et al., 1973] Meisner, C. W., Thorne, K. S., and Wheeler, J. A. (1973). *Gravitation.* W.H. Freeman and Company.

[Moffat, 2008] Moffat, J. W. (2008). *Reinventing Gravity, A Physicist Goes Beyond Einstein.* Thomas Allen Publishers.

[Penrose, 2005] Penrose, R. (2005). *The Road to Reality A Complete Guide to the Laws of the Universe.* Vintage.

[Roos, 2015] Roos, M. (2015). *Introduction to Cosmology.* John Wiley and Sons, Inc.

[Ryden, 2003] Ryden, B. (2003). *Introduction to Cosmology.* Cambridge University Press.

[Schwartz, 2014] Schwartz, M. D. (2014). *Quantum Field Theory and the Standard Model.* Cambridge University Press.

[Spiegel, 1973] Spiegel, M. (1973). *Mathematical Handbook of Formulas and Tables.* Schaum's outline series. McGraw-Hill.

[Susskind and Friedman, 2014] Susskind, L. and Friedman, A. (2014). *Quantum Mechanics, The Theoretical Minimum.* Basic Books.

[Susskind and Friedman, 2017] Susskind, L. and Friedman, A. (2017). *Special Relativity and Classical Field Theory, The Theoretical Minimum.* Basic Books.

[Susskind and Hrabovsky, 2014] Susskind, L. and Hrabovsky, G. (2014). *Classical Mechanics, The Theoretical Minimum.* Penguin Group.

[Tong, 2021] Tong, D. (2021). Particle physics: Cern lectures. Last accessed on 28 May 2024.

[Wang et al., 2017] Wang, Q., Zhu, Z., and Unruh, W. G. (2017). How the huge energy of quantum vacuum gravitates to drive the slow accelerating expansion of the universe. *Phys. Rev. D*, 95:103504.

[Weinberg, 1972] Weinberg, S. (1972). *Gravitation and Cosmology.* John Wiley and Sons, Inc.

[Weinberg, 1983] Weinberg, S. (1983). *The First Three Minutes, A Modern View of the Origin of the Universe.* Bantam Books, Inc.

[Weinberg, 2015] Weinberg, S. (2015). *Cosmology.* Oxford University Press.

[Wikipedia, 2022a] Wikipedia (2022a). Black hole. https://en.m.wikipedia.org/wiki/Black_hole. Last accessed on 18 June 2022.

[Wikipedia, 2022b] Wikipedia (2022b). Cosmology. https://en.m.wikipedia.org/wiki/Cosmology. Last accessed on 18 June 2022.

[Wikipedia, 2022c] Wikipedia (2022c). Quantum field theory. https://en.m. wikipedia.org/wiki/Quantum_field_theory. Last accessed on 10 June 2022.

[Wikipedia, 2022d] Wikipedia (2022d). Quantum mechanics. https://en.m. wikipedia.org/wiki/Quantum_mechanics. Last accessed on 10 June 2022.

[Wikipedia, 2022e] Wikipedia (2022e). Standard model. https://en.m.wikipedia. org/wiki/Standard_Model. Last accessed on 18 June 2022.

[Wikipedia, 2022f] Wikipedia (2022f). String theory. https://en.m.wikipedia. org/wiki/String_theory. Last accessed on 18 June 2022.

[Wikipedia, 2022g] Wikipedia (2022g). Theory of relativity. https://en.m. wikipedia.org/wiki/Theory_of_relativity. Last accessed on 10 June 2022.

[Wikipedia, 2023] Wikipedia (2023). Loop quantum gravity. https://en.m. wikipedia.org/wiki/Loop_quantum_gravity. Last accessed on 02 February 2023.

[Wray, 2011] Wray, K. (2011). An introduction to string theory. http://math. berkeley.edu/~kwray/papers/string_theory.pdf. Last accessed on 19 June 2022.

[Zee, 2010] Zee, A. (2010). Quantum Field Theory in a Nutshell. Princeton University Press.

[Zee, 2023] Zee, A. (2023). *Quantum Field Theory, As Simply as Possible.* Princeton University Press.

[Zwiebach, 2009] Zwiebach, B. (2009). *A First Course in String Theory.* Cambridge University Press.

Index

www.ingramcontent.com/pod-product-compliance
Lightning Source LLC
Chambersburg PA
CBHW050627190326
41458CB00008B/2174